Nanotechnology and the Environment

Errata Sheet

ACS Symposium Series 890

Nanotechnology and the Environment: Applications and Implications

Barbara Karn, Tina Masciangioli, Wei-xian Zhang, Vicki Colvin, and Paul Alivisatos

On page 3 in Chapter 1"Overview of Environmental Applications and Implications: How does Nanotechnology Relate to the Environment? Or Why Are We Here" by Barbara Karn, some text was inadvertently omitted.

This language set the stage for using new technologies to help enable that "productive harmony." Focusing on the small of nanotechnology has the potential to help create the environments envisioned by the authors of this Act.

Nanotechnology and the Environment

On page 4 before the bullet list, the following paragraph is missing.

These issues all deal with sustainability, a broad goal that goes beyond environmental protection to include social and economic issues as well. It has been defined in several ways:

The bullet list is correct and following the bullet list copy should read

No matter which definition is used, reading vital signs of the planet tell us we need to change the path we are on in order to reach sustainability. Issues like the.... (Pick up rest of paragraph).

ACS SYMPOSIUM SERIES **890**

Nanotechnology and the Environment

Applications and Implications

Barbara Karn, Editor
U.S. Environmental Protection Agency

Tina Masciangioli, Editor
National Academy of Sciences

Wei-xian Zhang, Editor
Lehigh University

Vicki Colvin, Editor
Rice University

Paul Alivisatos, Editor
University of California at Berkeley

Sponsored by the ACS Divisions
Industrial and Engineering Chemistry, Inc.,
Environmental Chemistry, Inc., Geochemistry, Inc., and
Polymeric Chemistry: Science and Engineering, Inc.

American Chemical Society, Washington, DC

Library of Congress Cataloging-in-Publication Data

Nanotechnology and the Environment : applications and implications / Barbara Karn ....
[et al.], editors.

    p. cm.—(ACS symposium series ; 890)

    Includes bibliographical references and index.

    ISBN 0–8412–3877–4 (alk. paper)

    1. Nanotechnology—Congresses. 2. Nanotechnology—Environmental aspects—
Congresses. 3.Nanoparticles—Environmental aspects—Congresses.

    I. Karn, Barbara, 1942- II. Series.

T174.7.N3733   2004
620′.5—dc22   2004050263

The paper used in this publication meets the minimum requirements of American National Standard for Information Sciences—Permanence of Paper for Printed Library Materials, ANSI Z39.48–1984.

PRINTED IN THE UNITED STATES OF AMERICA

Foreword

The ACS Symposium Series was first published in 1974 to provide a mechanism for publishing symposia quickly in book form. The purpose of the series is to publish timely, comprehensive books developed from ACS sponsored symposia based on current scientific research. Occasionally, books are developed from symposia sponsored by other organizations when the topic is of keen interest to the chemistry audience.

Before agreeing to publish a book, the proposed table of contents is reviewed for appropriate and comprehensive coverage and for interest to the audience. Some papers may be excluded to better focus the book; others may be added to provide comprehensiveness. When appropriate, overview or introductory chapters are added. Drafts of chapters are peer-reviewed prior to final acceptance or rejection, and manuscripts are prepared in camera-ready format.

As a rule, only original research papers and original review papers are included in the volumes. Verbatim reproductions of previously published papers are not accepted.

ACS Books Department

Contents

Environmental Implications: Nanoparticle Geochemistry in Water and Air
(Alexandra Navrotsky, Editor)

Environmental Implications: Metrology for Nanosized Materials
(Clayton Teague and Barbara Karn, Editors)

Environmental Applications: Sensors and Sensor Systems (Susan L. Rose-Pehrsson and Pehr E. Pehrsson, Editors)

Environmental Applications: Treatment and Remediation Using Nanotechnology
(Daniel Strongin and Wei-Xian Zhang, Editors)

Environmental Applications: Nanocatalysts for Environmental Technology
(Sarah C. Larsen, Editor)

Environmental Applications: Environmental Benign Nanomanufacturing
(Ajay P. Malshe and K. P. Rajurkar, Editors)

Environmental Applications: Nanotechnology–Enabled Green Energy and Power
(Debra R. Rolison, Editor)

Indexes

Preface

When we first envisioned an ACS symposium on nanotechnology and the environment, we realized that there was very little precedent for examining the relationship of this new technology and its impacts, either useful or harmful, on the environment. The National Nanotechnology Initiative had addressed environment as part of Societal Impacts, but the majority of work that could be described as environmental nanotechnology was directed toward natural nanoparticles in the environment—nanogeochemistry and atmospheric a erosols. O ur t hrust was to bring together nanotechnology research that contributed to enhanced environmental protection directed toward human activities, in addition to helping define the problems and processes that might occur in the natural environment. With the human aspects of new technologies in mind, we organized the symposium around two broad themes: the applications of nanotechnology to the environment and implications of nanotechnology on the environment. The applications were those aspects of nanotechnology that could be useful in dealing with current environmental problems and preventing future problem. They included treatment and remediation of existing pollutants, nanocatalysts for more selective and efficient reactions, nanotechnology-enabled green energy, metrology to measure these small materials, nanotechnology-enabled sensors for substances of environmental interest, and environmentally benign manufacturing of nanomaterials. Implications included toxicology and biointeractions of nanomaterials and nanoparticle geochemistry in water and air.

Although the symposium was organized as part of the American Chemical Society (ACS) meeting, our presenters came from a wide spectrum of disciplines representative of the interdisciplinary nature of nanotechnology. Materials engineering, chemistry, physics, journalism, mechanical engineering, nuclear engineering, toxicology, electrical engineering, chemical engineering, as well as civil and environmental engineering were represented. Rather than sessions directed toward a subset of one discipline, our sessions were organized around the potential products of research or the problems that might arise because of the size scale and technology.

In addition to diverse backgrounds, speakers came from diverse geographical areas and institutions. All presenters came from U.S. institutions, with Pennsylvania, California, Washington, Texas, and New York having the most representation. About one-quarter of the speakers were women.

Participants in the symposium were charged to think about three questions as they listened to the talks: What is the potential of this work to address environmental issues? How can the potential technology prevent environmental problems? Will the application have unintentional environmental implications? With these questions and the research presented, we hoped to build a community of researchers who are both environmentally conscious and to gain an environmental conscience so they can think of both the applications and implications of their own work to the environment. We hope this book will provide a framework and useful information for addressing nanotechnology and the environment.

We gratefully acknowledge the financial support of the ACS Division of Industrial and Engineering Chemistry, Inc. In addition we thank our fellow editors, the many authors, and the reviewers who contributed to this publication.

Barbara Karn
U.S. Environmental Protection Agency
1200 Pennsylvania Avenue, 8722F
Washington, DC 20460
karn.barbara@epamail.epa.gov

Tina Masciangioli
National Academy of Science
500 Fifth Street, NW, W629
Washington, DC 20001
tmasciangoli@nas.edu

Vicki Colvin
Department of Chemistry, MS 60
Rice University
Houston, TX 77005
colvin@rice.edu

Wei-xian Zhang
Department of Civil and Environmental Engineering
Lehigh University
13 East Packer Avenue
Bethlehem, PA 18015–3045
wez3@lehigh.edu

Paul Alivisatos
Department of Chemistry
University of California at Berkeley
Berkeley, CA 94720–1460
alivis@uclink4.berkeley.edu

Pertinent Reports and Websites

Royal Society Report ("Nanoscience and nanotechnologies: opportunities and uncertainties"), July 2004
(http://www.nanotec.org.uk/finalReport.htm)
Report was published by the Royal Society and the Royal Academy of Engineering after being commissioned by the Government in 2003. The report recognizes the many current and potential benefits of nanotechnology while indicating the need for public debate about their development. It also addresses research and regulation as they pertain to immediate safety with regard to health and the environment.

Societal Implications of Nanotechnology, publications from 1999-2004
(http://nano.gov/html/res/home_res.html)
National Nanotechnology Initiative (NNI) website includes reports on nanotechnology research and the possible implications for society. Sources of the reports listed include the National Research Council, the National Science Foundation, and the Nanoscale Science, Engineering and Technology Subcommittee (NSET) of the NNI.

EPA Office of Research and Development, National Center for Environmental Research
http://www.epa.gov/ncer
Nanotechnology and the environment grants (44) are listed on this web page. In addition, reports of Grantees' meetings on nanotechnology research, the Grand Challenge Research planning for nanotechnology and the environment report, papers and presentations from the Interagency meeting on nanotechnology and the environment are posted.

Swiss Re: Nanotechnology Small Matter, Many Unknowns, 2004
(http://www.swissre.com/INTERNET/pwswpspr.nsf/vwRobotCrawlLU/
LCLN-5Z5LUK?OpenDocument&RobotCrawl=1) or
http://www.swissre.com

The Swiss Reinsurance Company compiled this report assessing the potential implications of nanotechnology with the input of their risk officers. The report focuses on human exposure routes and occupational hazards, threats to health and the environment, and regulation. Swiss Re goes on to analyze these implications in the context of the insurance industry.

International Dialogue on Responsible Nanotechnology, June, 2004
(http://www.nsf.gov/home/crssprgm/nano/dialog.htm)
A meeting of government representatives from 26 countries on the societal implications of nanotechnology held in Alexandria, Virginia. Issues addressed included current and potential regulation from each nation, research and development investment, environmental protection, and public knowledge and perception of nanotechnology.

Center for Biological and Environmental Nanotechnology
http://www.ruf.rice.edu/~cben/index.shtml
This NSF-sponsored research center at Rice University is dedicated to addressing the scientific, technological, environmental, human resource, commercialization, and societal barriers that hinder the transition from nanoscience to nanotechnology.

Chemical Industry R&D Roadmap for Nanomaterials by Design: From Fundamentals to Function, December, 2003
(www.chemicalvision2020.org/pdfs/nano_roadmap.pdf)
Prepared by Chemical Industry Vision2020 Technology Partnership and Energetics, Inc., and sponsored by the Department of Energy and NSET, this report emphasizes a solution-oriented approach to materials development, termed "Nanomaterials By Design." The report asserts that such an approach would allow material producers to focus on the requirements for specific applications as the primary drivers of the design process, accelerating development.

Nanotechnology and the Environment

Environmental Applications and Implications

Editor
Barbara Karn

Chapter 1

Overview of Environmental Applications and Implications. How Does Nanotechnology Relate to the Environment? Or Why Are We Here?

Barbara Karn

Office of Research and Development, National Center for Environmental Research, U.S. Environmental Protection Agency, 401 M Street, SW, Washington, DC 20460

Nanometer scale (1-100 nm) science, engineering, and technology (collectively known as nanotechnology) has the potential to substantially enhance environmental quality and sustainability through utilizing nanomaterials to improve detection and sensing techniques for biological and chemical toxins, removal and destruction of the finest contaminants from air, water and soil; and the discovery of new " green" industrial processes that reduce energy and resource use and generation of waste products. In this paper, the relationship of this new technology to the environment and its potential for environmental protection are reviewed.

Introduction

What does nanotechnology have to do with the environment?...and why is an agency whose mission is "to protect human health and safeguard the natural environment-air, water, land - upon which life depends" involved in organizing a meeting of scientists and engineers to examine the relationship both positive and potentially negative of a new technology with its mission? This paper will address the latter question by answering the former.

In 1970 the National Environmental Policy Act declared that it is the "continuing policy of the Federal Government, in cooperation with State and local governments, and other concerned public and private organizations, to use all practicable means and measures, including financial and technical assistance, in a manner calculated to foster and promote the general welfare, to create and maintain conditions under which man and nature can exist in productive harmony, and fulfill the social, economic, and other requirements of present and future generations of Americans. The Congress, recognizing the profound impact of man's activity on the interrelations of all components of the natural environment, particularly the profound influences of population growth, high-density urbanization, industrial expansion, resource exploitation, and new and expanding technological advances and recognizing further the critical importance of restoring and maintaining environmental quality to the overall welfare and development of man.

Today's manufacturing methods "move atoms in great thundering herds." (Ralph Merkle, Xerox). We get what we want by removing matter in a top-down approach. We grind, we cast, we sinter, we saw. We irreparably change the landscape when we extract natural resources from the ground. We then use large amounts of energy to extract and cast materials at high temperatures often merely to get a simple product like an aluminum fence post.

Manufacturing at the nanoscale is different. Its basis is construction using molecular building blocks and precise positional control. This molecule-by-molecule control is the potential basis of a manufacturing technology that is cleaner and more efficient than anything we know today. It is a fundamentally different way of processing matter to make products for human wants and needs.

In Nature manufacturing begins at the bottom and builds up materials, using the information in DNA to form RNA which assembles proteins inside cells. These proteins are enzymes that catalyze the building of cellular materials and enable a hierarchical structure ranging from cells to tissues to organs and, in the end, the organisms that are the life on the planet. Nature does most of this building by combining "natural," simple and abundant elements (carbon, hydrogen, oxygen, nitrogen) near room temperature, using tiny machines for assembling (ribosomes), in a non-toxic solvent (water), using efficient and highly selective catalysts (proteins) with the end of life and/or recycling accounted for

4

(bacterial decay and remineralization). If we cannot achieve this full process with nanotechnology, at the least, it can be a goal for our human industries.

If we continue to produce goods as we have in the past, we will not like the result for the environment. However, we do not want to give up our lifestyle. The answer is not to give people fewer and poorer quality goods but rather how to provide for the needs of people with less impact to the environment. We cannot just get better at our current processes, but we need to develop new processes. At the moment, the best opportunity for getting better at making things lies in the field of nanotechnology. This is an opportunity to take what we are doing in nanotechnology to change the course of our human activities toward a more sustainable path.

Sustainability

In hindsight, what kind of results have the human industries of the current Industrial Revolution brought us--as "getting and spending we lay waste our powers...?" Do we live in the kind of environment we want? What do we think of when we picture our ideal environment? alpine vistas of snow capped peaks viewed through clear air, an urban townhouse on a quiet tree-lined street with friendly neighbors walking by, dolphins surfacing across a clear bay, or a rare bird visiting our garden? We see no environmental problems in any of these visions. We do not envision the carbon dioxide from our power plants and industrial activities that increases the potential for global climate change nor the unclean water that plagues most of the world's population nor the toxic wastes that must be disposed of as a result of our industrial production. We did not envision an environment of urban poverty and decay nor large-scale epidemics nor loss of natural systems and species diversity.

• "Development that meets the needs of the present without compromising the ability of future generations to meet their needs" [1]

• "The reconciliation of society's developmental goals with the planet's environmental limits over the long term" [2]

• "Meeting fundamental human needs while preserving the life-support systems of planet Earth" [3]

world population passing 6 billion, groundwater depletion continuing, vehicle production increasing, paper piling up, the number of wars increasing, urban populations rising, fish harvests decreasing, grain harvest falling, meat production increasing are signals that we are on a non-sustainable course. The negative signals are not yet compensated for by some positive signs such as environmental treaties gaining ground, fertilizer use moving down, compact fluorescents sales increasing and the use of information technology and the internet accelerating.[4].

The issues in sustainability can be broken into five categories :

global climate change;

depletion of natural resources, including water, forest products, minerals, and fossil fuels;

population problems, including infectious disease;

urbanization and social disintegration, income gaps; and

environmental degradation, including pollution, threatened habitats, and loss of biodiversity. Nanotechnology can help with all these sustainability issues. For example, the potential for global climate change can be limited as nanotechnology produces lighter weight materials that use less energy in their production and lighten our transportation products thereby burning less fossil fuel and emitting fewer greenhouse gases. Efficient electronics which use less electrical energy, more efficient product manufacturing using less production energy, nanotechnology-enabled alternative energy sources, and cleaner burning fuels due to nanotechnology applied to more efficient filters for better prefiltration are other examples of how global climate change may be lessened using nanotechnology.

In the area of water quantity and quality nanotechnology can improve water filtration systems for drinking water purification and waste removal, and form the basis for sensors to detect both chemical and biological water pollutants. Nanotechnology-enabled sensors can help manage forest ecosystems, while materials are saved through dematerialization—less use of materials as nanotech enables production of smaller products and less waste by building from the bottom up. Fuels can be saved through nanotechnologies that enable moving to alternative fuels such as solar, hydrogen, and enabling more efficient use of petroleum in materials manufacturing.Population-related problems addressed by nanotechnologies could include better materials for urban infrastructures—e.g., smart sewer pipes, distributed water treatment, less expensive food crops grown with more efficient nanosized fertilizers and pesticides, eliminating health problems of epidemic proportions by increasing targeted drug delivery in the body, bottom up synthesis of medicines and other useful bioactive compounds, better analytical tools for detecting, preventing diagnosing and curing diseases.

Nanotechnology addresses environmental problems by more efficient chemical manufacturing with less toxic wastes and different processes leading to fewer by products, use of less toxic starting materials; better ways to detect pollutants in air, water, and on land; improved treatment and remediation using nanoparticles with high surface area and active sites; habitat protection increased through freed-up capital from other forms of waste treatment; improved ecosystem research tools like nano-transmittors to track animal movements; and improved, inexpensive sensors to detect subtle ecosystem changes. As nanotechnology converges with biotechnology, information technology, and cognitive science, the "nano, bio, info, cogno" or "NBIC" convergence, the combined technologies will provide an even stronger foundation for sustainability than nanotechnology alone.

Implications of Nanotechnology

However, as in any 'new technology, nanotechnology can allows the possibility for abuse that can harm the environment and sustainability. For example, there could be development of new disease organisms, nano-scale agents for crop destruction, forest cover removal (a new Agent Orange), or use in "area denial," a different kind of land mine. Whether it is used for society's benefit or detriment, as the Foresight Institute has acknowledged, nanotechnology will be developed regardless of efforts to suppress it. Nevertheless, proper regulation and "the weapon of openness" can control risks of nanotechnology's improper use, and regulating nanotechnology will be a process, not an event.

Regulatory agencies will have to make some decisions about nanotech such as: is a nanomaterial a drug or a device? Is it a chemical or a biological agent? How much risk (toxicity and exposure) is associated with nanomaterials? How much does size matter? Is the current regulatory structure adequate, or is new legislation needed to ensure the protection of the environment and human health? As this new technology is developed, regulatory agencies will have to concurrently examine these questions and others in addressing any implications of nanotechnology.

Conclusion

Nanotechnology enables a powerful new direction for our current industries and human activities. Research informed about environmental applications and implications of nanotechnology can help bring about a future environment that is different from and better than the current environment. This research in environmental nanotechnology must maintain a commitment to rigor with applications related to sustainability and awareness about and prevention of possible harmful consequences, seeking advice on making sustainable choices.

References

[1] Bruntland, G. (ed.), (1987), "Our Common Future: The World Commission on Environment and Development", Oxford, Oxford University Press.
[2] Board on Sustainable Development, Policy Division, National Research Council (1999), "Our Common Journey: A Transition Toward Sustainability" Washington, D.C. National Academy Press
[3] Kates, RW, et. al., (2001) *Science: 292 :* 641-642
[4] Worldwatch Institute (2003) "Vital Signs," Washington D.C.

Chapter 2

Environmental Technologies at the Nanometer-Scale

Tina M. Masciangioli[1,3] and Wei-xian Zhang[2]

[1]AAAS Environmental Science and Technology Policy Fellow, National Center for Environmental Research, U.S. Environmental Protection Agency, Washington, DC 20460
[2]Department of Civil and Environmental Engineering, Lehigh University, Bethlehem, PA 18015
[3]Current address: National Academy of Science, 500 Fifth Street, NW, W629, Washington, DC 20001

Nanometer scale (1-100 nm) science, engineering, and technology (collectively known as nanotechnology) has the potential to substantially enhance environmental quality and sustainability through utilizing nanomaterials to improve detection and sensing techniques for biological and chemical toxins, removal and destruction of the finest contaminants from air, water and soil; and the discovery of new " green" industrial processes that reduce energy and resource use and generation of waste products. In this work, promising environmental applications of nanotechnology for pollution prevention are reviewed.

Introduction

As we are faced with major environmental challenges in the 21st century, new approaches are required to maintain and improve soil, water, and air quality. As discussed previously,[1] we anticipate nanotechnology playing a key role in the enhancement of environmental technologies. Here we focus our attention on nanotechnology for pollution prevention -- the reduction or elimination of waste at the source. Pollution prevention efforts will likely be enabled by nanotechnology in the short-term through improvements in material properties, such as the use of nanoparticles as additives and fillers.

Although various fillers have been used for many years, such as the use of carbon particles (carbon black) as filler in tire rubber for about 100 years (for color and improved mechanical properties), or metal nanoparticles in glasses and ceramic glazes for centuries,[2] improvements in our understanding and control of the nanometer scale by way of new tools and synthetic techniques are leading to a revolution in new material properties.

Because so many fields, like energy, environment, transportation, housing, and health depend on new and better materials, advances made through nanotechnology are necessarily wide in their impact. Moving from fossil fuels as an energy source to solar energy (which would provide tremendous environmental benefits), for example, requires significant advancements in our ability to capture sunlight and convert it into electricity. This ability will come with improvements in photovoltaic, energy storage, and energy conduction materials. It is anticipated that nanotechnology will enable such improvements in the not too distance future.

Technologies

Lighting

High brightness light-emitting diodes (LEDs) constructed from nanostructured solid-state inorganic or organic materials have been referred to as the 'lighting of the Third Millennium for the entire world" by the Light up the World Foundation (www.lightuptheworld.org), founded by electrical engineer Dave Irvine-Halliday. These LEDs have been discussed extensively elsewhere,[3] but still deserve further mention here due to the their tremendous potentional for preventing pollution.

The material used in these high brightness LEDs is grown by nanotechnology based compound semiconductor epitaxial growth techniques -- from materials like AlInGaP and AlInGaN. Pervasive use of this technology could reduce energy consumption by an estimated 50% in the U.S. by 2020, reducing carbon emissions by 28 million metric tons per year, and at a cost savings of $100 billion over the period 2000-2020.[3, 4] Savings comes not only from the lower power usage, but also from how robust and long lasting these lights are.

White LED technology holds even greater potential for the developing world. The Light Up the World Foundation is now bringing white LED technology to the world's poorest, most remote villages to light homes, schools, and temples; places that now rely on kerosene wick lamps or candles as light sources. It has been found that a single 0.1-watt, white LED supplies enough light for a child to read by, thus making it possible to light up an entire village with less energy than that used by a single, conventional 100-watt light bulb.

Transportation

Nanotechnology may enable the production of automobiles in the future that are lighter, stronger, and less polluting, through the use of lightweight but durable clay nanoparticle, carbon nanotube, and biofiber nanocomposites[5].

Such composite materials have gainied great interest from the automotive industry, especially since the European Union (EU) directive was made that by 2015 vehicles must be made of 95% recyclable materials.[6] Using biofiber-nanocomposites will be especially helpful for satisfying this stringent environmental criteria while providing the ability to reduce vehicle weight. Reducing weight is one of the main reasons for moving to composite materials for automobile parts but the use of glass or carbon fiber polymer composites are more difficult to receycle.

Recently[7], the life cycle economic and environmental implications of using clay-polypropylene nanocomposite or aluminum in automobiles instead of steel in light-duty vehicle body parts were evaluated. Substantial benefits from reducing energy use and environmental discharges were demonstrated. For example, substituting the nanocomposite material or alumninum for steel in one years's fleet of vehicles in the U.S., resulted in an energy savings of 50-240 thousand tera joules, a reduction in 4-16 million tons of $CO2$ equivalents of greenhouse gases released, a savings of 5-6 million tons of ore, and as much as 7 fewer occupational fatalities.

The highly praised carbon nanotube[8] provides other environmental benefits when incorporated into composite materials. By themselves carbon nanotubes are many times stronger than steel at a sixth of the weight, but can also be conductive like steel. When present in plastics for automobiles carbon nanotubes enable use of electrostatic painting.[9] Electrostatic painting of automobiles offers approximately 3 times higher paint transfer efficiency over regular spray painting and reduces paint usage and VOC emission in the automobile manufacturing process. Carbon nanotube and other composite materials, however, are currently very expensive. Researchers[10] are thus working hard on ways to improve production yields and lower the costs of these materials for their use is in large quantity.

Renewable Energy

One of the most promising application areas of nanotechnology is for capturing solar energy. Pioneering research from the 1990's utizing dye sensitized TiO_2 is now being commercialized by Konarka Technologies, Inc. (www.konarkatech.com) for the production of lightweight, low cost plastic solar cells. In addition to the great benefit of using a renewable energy source, these plastic cells also do not require high temperatures or other stringent manufacturing conditions as in silicon technology – thus, making it a more "green" manufacturing process. The plastic cells are also suitable for a use in wide range of places, including rooftops, window blinds, cars, or even clothing. These plastic cells, however, are less efficient than silicon and thus, substantial improvements are still needed to both lower cost of solar cells, and improve their power efficiencies before there will be any wide-spread use.

Recently, advances have been made in a new type of "paint-on" solar cell[11] that also shows promise for relatively low fabrication cost and efficiencies comparable to those of high-end silicon based technology. These photovoltaic devices are organic-inorganic hybrids, based on a mixture of a the conjugated polymer poly-3(hexylthiophene) and 7-nanometer by 60-nanometer CdSe nanorods. The nanorods enable tuning of the absorption spectra, to capture a greater range of solar light. They were assembled from solution with an external quantum efficiency of over 54% and a monochromatic power conversion efficiency of 6.9% under 0.1 milliwatt per square centimeter illumination at 515 nanometers. Under Air Mass (A.M.) 1.5 Global solar conditions, the researchers obtained a power conversion efficiency of 1.7%. As discussed above, solar cells made entirely from polymers have the potential to provide low-cost, ultra-light weight, and flexible cells, with a wide range of potential applications.

Other researchers are exploring ways to create solar cells inspired by or utilizing natural systems. Such as, at the NASA funded Center for Cell Mimetic Space Exploration at UCLA (http://www.cmise.ucla.edu/cmise_energetics.htm), where researchers hope to create a biosolar cell that is capable of directly converting optical energy into electricity using two proteins, bacteriorhodopsin and cytochrome oxidase. Together, these proteins can transform light into a proton gradient and then convert it into an electromotive force. The estimated power density of this biosolar cell exceeds 435 W/kg, while today's solar cell can only produce 80 – 140 W/kg. In related research, scientists at the University of Arizona[12] are exploring ways to carry out photosynthesis artificially through improved understanding of natural systems on the nanometer scale. Recently, this group demonstrated that a light-powered artificial photosynthetic reaction centre can be coupled to endergonic transmembrane Ca^{2+} transport by a redox loop to store energy as a combination of electrical and chemical potential. Although their work is biomimetic only in the broadest sense, the system they created illustrates how various elements of biological energy transduction may be combined in nanoscale synthetic constructs that perform interesting and potentially useful functions.

Potential Problems

As with any new technology or chemical substance, there is the potential for harm to human health and the environment from nanotechnology. However, there are mechanisms in place to help avoid such problems. Depending on their intended uses, new chemical substances are reviewed for safety and efficacy under federal statutes, such as the Federal Insecticide, Fungicide, and Rodenticide Act (FIFRA), the Federal Food, Drug, and Cosmetic Act (FFDCA), and the Toxic Substances Control Act (TSCA). Whether or not these statutes are truly effective for nanomaterials probably could and should be debated. Aside from this type of federal regulatory oversight, however, there are efforts taking place in the government to specifically address concerns about nanotechnology through scientific research. For example, the National Center for Environmental Research at the U.S. EPA announced that it will fund research to explore the health and environmental effects of manufactured nanomaterials (www.epa.gov/ncer).

Careful consideration of both the environmental and economic benefits of new nanotechnologies must also be evaluated. While nanotechnology presents new ways to address environmental problems and even prevent pollution, in order for such technologies to be adopted they will need to be cost effective. This also holds true for creating new materials for commercial use. Much of our material and resource use, and many of our current manufacturing methods have not changed in decades. For example, as pointed out in the life cycle report mentioned earlier[7], there is a well established history of producing a quality product from steel. It was found that even without going to a nanocomposite material, aluminum would offer great weight reduction and environmental savings as compared to steel for vehicle manufacuterers. However, the automobile industry has not started substituting aluminum for steel in any substantial way.

Conclusions

Through new understanding and manipulation of the nanometer length scale, it is possible to develop significantly improved properties of materials and devices. These improvements could lead to great benefits for human health and the environment. Lower energy requirements lead to reductions in emission of pollutants. Using less material while achieving enhanced properties leads to reductions in waste. However, creating more environmentally benign products alone cannot solve pollution problems. The use of high brightness LEDs, for example, may have substantial environmental benefits in terms of reducing energy consumption, but there may be large environmental costs associated with the production of the LED materials and with disposal of the materials after use. As nanotechnology research and development progresses, it will be necessary to look at the total life-cycle implications and compare to existing well established materials, to assess the true environmental benefits of these new materials.

Acknowledgements

TMM wishes to thank EPA colleagues Dr. B. Karn, Dr. D. Bauer, Dr. N. Savage, as well as AAAS and U.S.EPA for support during AAAS S&T policy fellowship.

References

1.   Masciangioli, T. and W.-x. Zhang, *Environmental Technologies at the Nanoscale.* Enivron. Sci. Technol., 2003. **37**(5): p. 102A-108A.
2.   Padovani, S., et al., *Copper in glazes of Renaissance luster pottery: Nanoparticles, ions, and local environment.* J. Applied Phys., 2003. **93**(12): p. 10058-10063.
3.   Talbot, D., *LEDS vs. The Lightbulb.* Technology Review, 2003. **106**(4): p. 30-36.
4.   Bergh, A., et al., *The Promise and Challenge of Solid-State Lighting.* Physics Today, 2001. **54**(12): p. 42-.
5.   Mohanty, A.K., L.T. Drzal, and M. Misra, *Nano Reinforcements of Bio-based Polymers -- The Hope and the Reality.* Polymeric Materials: Science & Engineering, 2003. **88**(American Chemical Society Spring Meeting): p. 60-61.
6.   Marsh, G., *Next step for automotive materials.* Materials Today, 2003. **6**(4): p. 36-43.
7.   Lloyd, S.M. and L.B. Lave, *Life Cycle Economic and Environmental Implications of Using Nanocomposites in Automobiles.* Environ Sci Technol, 2003. **37**: p. 3458-3466.
8.   Baughman, R.H., A.A. Zakhidov, and W.A. de Heer, *Carbon Nanotubes--the Route Toward Applications.* Science, 2002. **297**(5582): p. 787-792.
9.   Xiang, X.D., *Combinatorial Screening of High-Efficiency Catalysts for Large-Scale Production of Pyrolytic Carbon Nanotubes*, E. SBIR, Editor. 2002, Intematix Corporation: Washington.
10.   *Carbon nanotubes for static dissipation.* Plastics,Additives and Compounding, 2001. **3**(9): p. 20-22.
11.   Huynh, W.U., J.J. Dittmer, and A.P. Alivisatos, *Hybrid Nanorod-Polymer Solar Cells.* Science, 2002. **295**(5564): p. 2425-2427.
12.   Bennett, I.M., et al., *Active transport of Ca2+ by an artificial photosynthetic membrane.* Nature, 2002. **420**: p. 398 - 401.

Chapter 3

Reassessing Risk Assessment

Douglas Mulhall

Our Molecular Future, 2022 Cliff Drive, #343, Santa Barbara, CA 93109

Introduction and Background

By revising risk assessment to account for new scientific discoveries, industry and regulators may be able to collaborate in areas where they might otherwise be at odds over perceived risks and benefits of technology convergence.

For example, according to some environmental scientists, new "nanomaterials" are too risky to manufacture until we study them further(*1*). Other experts disagree and say that potential benefits outweigh the risks.(*2*) In my book, *Our Molecular Future*(*3*), I argue that methods used by both sides to evaluate such risks are becoming obsolete, and that only by overhauling them is it possible to accurately understand risks and benefits.

The good news is that tools to revitalize risk assessment are being developed and deployed. Such tools may also let industry adapt rapidly to newly discovered natural environmental threats. This rapid adaptation may let us *match nature's complexity*.

The author has pioneered environmental risk assessment methods(*4*) used by chemicals companies and government agencies. Based on that experience, and using research done for my book, I identified categories of technological advances and discoveries that may alter the basis for risk assessment. They include:

Emergence of Enhanced Human and Machine Intelligence.

Hans Moravec, head of Carnegie Mellon's Robotics Institute, has shown that the capacity to process information is multiplying exponentially, and the *rate* at which it is multiplying is increasing(*5*).

For millennia, this exponential rate was imperceptible in real time, because it took thousands, then hundreds, then tens of years for such capacity to multiply; from the abacus to the microprocessor, and now the nanoprocessor.

Today, this exponential acceleration is becoming perceptible. For example, it has enabled the rapid transition from desktop printing to three-dimensional desktop manufacturing (6). Such machines are effectively eliminating the conventional factory.

In 2001, a computer used "genetic computing" (7) to build a thermostat and actuator that were superior to the counterparts designed by a human. The computer's programmers were unable to trace how the computer reached its conclusion. This is because genetic algorithms let computers solve problems in their own way.

Machines with such enhanced intelligence do certain things far faster and better than we do. Not everything, but many things. Stock brokers now use computers that forecast commodity markets more accurately than humans do(8). Satellites that repair themselves and make unilateral data transmission decisions are already in orbit(9).

Such merging of human intelligence with genetic algorithms and massive networks is being applied to modeling of, for example, climate change. Yet we are only beginning to apply them to real-time evaluation of risk-related phenomena such as hurricane, earthquake, and tsunami that are still unpredictable in the very short term.

Many risk assessment methods still implicitly assume that evolution of intelligence will proceed in the same way that it has over the past few thousand years—gradually and without perceptible change in the foreseeable future. For example, it is still assumed that self-assembly and disassembly by complex, intelligent products is science fiction that is not achievable in the foreseeable future. It is assumed that intelligent structures that adapt rapidly to climate changes, earthquakes, and hurricanes are too expensive to be broadly adopted. Yet the aforementioned advances suggest that such assumptions are rapidly becoming invalid.

Technologies That Merge With The Natural Environment

Ray Kurzweil, who pioneered technologies such as the flatbed scanner, argues that "technology is a continuation of evolution by other means"(10). This implies that our technologies are becoming an integral part of the ecology. What are the physical manifestations of this?

Smart Dust(11) comprises a massive array of micromachines made of components that ride on air or water currents, undetected to the human eye. Each expendable machine can have a camera, communications device, and sensors for chemicals, temperature, and sound. It can serve as the eyes, ears, nose, and guidance mechanism for everyone from soldiers to tornado chasers. It is cheap to manufacture, and its prototype exists today. It forms part of an array that delivers information to one or hundreds of computers in one or many locations. It will soon be in our environment in the trillions, delivering real-time information about earth, air and water.

Such intelligent particles are also gaining the capacity to *self assemble*. Several universities have pioneered self-assembling photovoltaic materials that generate and conduct an electric current(*12*). These materials can be painted onto surfaces, thus eliminating need for solar panels.

When we combine self-assembly with intelligent sensing at the nanometer scale, and manufacture trillions of units, we see that technology becomes an integral part of the ecology instead of just impacting it. We may therefore have difficulty to distinguish between what has been manufactured and what has evolved "naturally".

This nanoscale level of incursion into the ecology by smart machines suggests emergence of an *intelligent environment* that is able to sense virtually every part of the ecology, from the epicenter of earthquakes to the heart of a hurricane and the heartbeat of every species, then interpret the data. Just as nature has its own intelligence by acting to balance environmental changes, so we are creating an intelligent human-built environment, not just alongside that, but as part of it.

New Evidence of Punctuated Equilibrium

The theory of "punctuated equilibrium(*13*) was first proposed in 1972 by Niles Eldredge and Harvard evolutionary biologist Stephen J. Gould. This holds that evolutionary changes occur relatively rapidly in comparatively brief periods of environmental stress, separated by longer periods of evolutionary stability. After many years of skepticism, their theory is gaining acceptance because proof is emerging.

Intelligent tools such as those described earlier, are helping us to discover that the natural ecology undergoes periodic large scale and localized instability that threatens our society, and not just in the extended timeframes that we used to think.

In 1994, Comet Shoemaker-Levy 9 (SL9) hit Jupiter(*14*), blasting holes the size of Earth in its atmosphere. Had this hit the Earth, human life would have been virtually extinguished. Until then, it was thought that such disruptions happened only every few million years and that we'd have time to see them coming. SL-9 demolished this idea. It demonstrated that we live in a galaxy where life can be snuffed out on a planetary scale without warning.

Furthermore, scientists have found that smaller events have upset the ecology here on Earth. Ice core and tree ring records show that in the year 536 A.D. an unknown event triggered a catastrophic cooling of the northern hemisphere, resulting in years without summers that led to wholesale crop failures and starvation(*15*). Other ice core and tree ring samples from around the world show that naturally induced climate flips have occurred far more frequently and suddenly than we once thought(*16*).

At the regional scale, in 1700 a fracture at the Cascadia subduction zone produced a tsunami that scoured the Pacific coast for miles inland where many of our cities now stand(*17*). In 1958, a *fifteen hundred-foot high wave* swept away a forest after a mountain collapsed into Lituya Bay, Alaska(*18*).

At the other end of the scale, nanosized organisms a hundred times smaller than most bacteria have been discovered(19,20,21). For decades, researchers around the world have theorized that many prevalent diseases such as atherosclerosis are caused by infection(*22*).This was proven for stomach ulcers decades ago, but for other illnesses no one could find a culprit. Now it seems that we have one. Skeptics claim that nano organisms don't exist, but studies in 2002 by the Mayo Clinic and other well-known universities now confirm their existence(*23*). Furthermore, there seems to be a treatment(*24*). Most recently, in March of 2003, nano scale micelle-like entities were found in large numbers in polluted river water(*25*), suggesting that a world of nano scale life forms may lie beneath the one that we are used to dealing with.

Such nano and macro scale phenomena may pose big threats, yet government agencies and insurers know of precious few defenses against them, and moreover they often do not have adequate methods for detecting them.

Thus, punctuated equilibrium is not part of the environmental risk assessment framework right now, and as such, a chunk of the ecological equation is missing.

This is especially true when considering the environmental risks and benefits posed by nanotechnologies. Such technologies may at once be driving the next "punctuation" in evolution, yet give us the tools to protect ourselves from conventional and newly discovered natural threats. That is the new paradox.

Discussion

Most of our agricultural, medical, and housing systems struggle to respond to the natural environment's complexity. Right now, many of our technologies are note very good at it. For example, we use antibiotics to cure infections, but they lose their potency when the environment that they work in adapts to them.

Yet this imbalance may shift. Technology convergence may let us do something that we have only dreamed of until now: <u>Adapt rapidly to nature's complexity.</u>

For example, we may be able to adapt rapidly to big natural onslaughts. We may build earthquake-proof and storm-proof structures based on newly discovered, super-strong, nanostructured aerogels that slash the costs and risks of building in risky zones. In medicine, drugs may be so precise that they backfire only occasionally instead of generating widespread immune responses as they do now.

This ability to adapt rapidly to nature's complexity may allow the institutions that underpin our economy, such as insurers for example, to redefine the boundaries of risk assessment for everything from property protection to disease vectors.

What to Do

Intelligent systems along with newly discovered natural threats are the elephants in the room of risk assessment. They must be investigated with a view to developing adaptive technologies such as those described in this paper. Yet this is not occurring among many of the institutions responsible for risk assessment.

For example, although nanoscale organisms have been identified in geological formations and the human body since the early1990's, neither the USEPA or insurance companies seem to examine the implications for human or natural ecology. Yet NASA is extensively studying it, as are Universities of Texas, McGill Canada, Regensburg and Hamburg Germany, Kuopio Finland, Melbourne Australia, and many others.

At the macro end of the scale, neither EPA nor insurance companies have yet examine newly discovered anomalies such as the naturally induced climate flip of circa 536 A.D., or the west coat tsunami of 1700. These might disrupt natural ecosystems and bankrupt insurers if they recur. Furthermore, no government agency or insurance company examines how nanotechnologies might be used to adapt to such phenomena.

I therefore suggest that as a first small step, EPA, FEMA, and selected insurers sponsor a forum: "<u>Match Nature's Complexity: Using technology convergence to adapt to newly discovered natural threats</u>". The goals of this forum would be:

--Summarize newly understood natural threats at the nano and macro scale,
--List advanced technologies that may help us adapt to them,
--Determine how risk assessment methods might be revised accordingly.

The forum might bring together experts who have never met as a group, such as: scientists who have developed aerogels and intelligent materials that can be used to adapt structures to climatic and seismic threats; researchers who discovered nanobacteria and developed methods to reverse nanobacterial infections; and scientists who have uncovered new naturally induced threats that warrant a new type of defense.

Notes

1. *The Big Down, Atomtech Technologies Converging at the Nano Scale*, ETC Group, January 2003. http://www.etcgroup.org/documents/TheBigDown.pdf
2. Updates on the nanoethics debates, including rebuttals to the ETC position are found in the December and February newsletters of CMP-Cientifica. http://www.cmp-cientifica.com/cientifica/frameworks/generic/public_users/tnt_weekly/subscribe.htm
3. Mulhall, Douglas, *Our Molecular Future - How Nanotechnology, Robotics, Genetics, and Artificial Intelligence Will Transform Our World.* Amherst New York, Prometheus Books 2002.
4. Michael Braungart, Justus Engelfried, Douglas Mulhall, "Criteria for Sustainable Development of Products and Production," Fresenius Environmental Bulletin 2: 70-77 Basel, Switzerland, 1993. Also, Mulhall, Douglas *Tools For Adapting To Big Ecosystem Changes*, Futures Research Quarterly, Vol 16 Number 3 49-61
5. Moravec, Hans, *Robot*, Oxford, 1998, Chapter 3: Power and Presence, page 60. "The number of MIPS in $1,000 of computer from 1900 to the present" http://www.frc.ri.cmu.edu/~hpm/book98/fig.ch3/p060.html [accessed March 15, 2003]
6. *Personal Fabrication on Demand,* Wired Magazine 9, no. 4 (April 9, 2001)
7. John R. Koza (Stanford University), Forrest H Bennett III (Genetic Programming Inc.), David Andre (University of California Berkeley), Martin A. Keane (Econometrics Inc.), *Genetic Programming III: Darwinian Invention and Problem Solving* (Morgan Kaufmann, 1999) Chapter V. *Automated Synthesis of Analog Electrical Circuits,*
8. Cliff, Dave, *Artificial Trading Agents for Online Auction Marketplaces*, HP Labs, Bristol http://www-uk.hpl.hp.com/people/dave_cliff/traders.htm [accessed March 15, 2003]
9. *Satellite Trio To Test Artificial Intelligence Software,* AviationNow.com, Aviation Week (May 30, 2001) [online], http://www.aviationnow.com/avnow/news/channel_space.jsp?view=story&id=news/ssat0530.xml [accessed March 15, 2003].

10. Kurzweil, Ray, *Are We Spiritual Machines?* Chapter 10 *The Material World: "Is That All There Is?* Future Positive http://futurepositive.synearth.net/2002/06/20 June 18, 2002

11. "Smart Dust", DARPA research website describing project on micro air vehicles http://robotics.eecs.berkeley.edu/~pister/SmartDust/ [accessed March 15, 2003].

12. *Solar cells go organic,* The Economist, Technology Quarterly, July 20, 2002 http://www.economist.com/science/tq/displayStory.cfm?story_id=1176099 [accessed March 14, 2003]

13. Gould, Stephen Jay, *Darwinian Fundamentalism,* New York Review of Books, June 12, 1997 http://www.nybooks.com/articles/1151 [accessed March 17, 2003]

14. Bruton, Dan, *Frequently Asked Questions About the Impact of Comet Shoemaker-Levy 9 with Jupiter,* Institute for Scientific Computation (February 2, 1996) [online], http://www.isc.tamu.edu/~astro/sl9/cometfaq2.html#Q3.1 [accessed August 29, 2001]

15. Stothers, R.B. "Mystery Cloud of 536 A.D," *Science Frontiers* No. 33 (May-Jun 1984) (re-printed from Stothers, R.B.; "Mystery Cloud of AD 536," *Nature* 307 (1984):344 http://www.science-frontiers.com/sf033/sf033p19.htm [accessed August 12, 2001].

16. *The Two Mile Time Machine: Ice Core, Abrupt Climate Change, and Our Future* Richard B. Alley, Princeton, N.J.: Princeton University Press, 2000. Also, *A Slice Through Time: Dendrochronology and Precision Dating,* M.G.L. Baillie, London: Routledge, 1995.

17. Nelson et al, "Radiocarbon Evidence for Extensive Plate Boundary Rupture About 300 Years Ago at the Cascadia Subduction Zone," Nature 378 (23 November 1995).

18. *The 1958 Lituya Bay Tsunami,* University of Southern California Tsunami Research Group, [online], http://www.usc.edu/dept/tsunamis/alaska/1958/webpages/index.html [accessed March 17, 2003].

19. Kajander O,Ciftcioglu N,Miller-Hjelle M, Hjelle T. Nanobacteria: controversial pathogens in nephrolithiasis and polycystic kidney disease. Curr Opin Nephrol Hypertens, 2001:10:445-452.

20. Folk, Robert L., *Nanobacteria: surely not figments, but what under heaven are they?* Texas Natural Science Journal, March 4, 1997 http://naturalscience.com/ns/articles/01-03/ns_folk.html [accessed January 25, 2003].

21. Harald Huber, Michael J. Hohn, Reinhard Rachel, Tanja Fuchs, Verena C. Wimmer & Karl O. Stetter, A new phylum of Archaea represented by a nanosized hyperthermophilic symbiont, Nature 417, 63 - 67 May 02, 2002.

22. Mawhorter, Steven D., Lauer, Machael A. *Is Atherosclerosis an Infectious Disease?* Cleveland Clinic Journal of Medicine, Vol 68, number 5, May 2001.
23. Todd E. Rasmussen, Brenda L. Kirkland, Jon Charlesworth, George P. Rodgers, Sandra R. Severson, Jeri Rodgers, Robert L. Folk, Virginia M. Miller, *Electron Microscope and Immunological Evidence of Nanobacterial-Like Structures in Calcified Carotid Arteries, Aortic Aneurysms and Cardiac Valves* Mayo Clinic and Foundation, Rochester, Minnesota; University of Texas, Austin, March 6, 2002.
24. NanobacTX is described at www.nanobaclabs.com
25. Martin Kerner, Heinz Hohenberg, Siegmund Ertl, Marcus Reckermannk & Alejandro Spitzy, *Self-organization of dissolved organic matter to micelle-like microparticles in river water.* Nature Vol 422, 13 March, 2003.

Chapter 4

The Environmental Impact of Engineered Nanomaterials

Kristen M. Kulinowski and Vicki L. Colvin

Department of Chemistry, Center for Biological and Environmental Nanotechnology (CBEN), Rice University, MS-60, 6100 Main Street, Houston, TX 77005

Nanoparticles are important in natural environments due to their size, tunable properties and accessible surfaces and our control over these properties can be exploited to create or add value to a variety of technologies. Many consumer products that incorporate nanoparticles, such as sunscreens and clothing, are already in the marketplace, and the industry is growing fast. This book highlights also the many valuable environmental technologies that can come from the applications of unique nanomaterial properties. As this nascent technology area matures, the debate about the whether the unknown risks of nanomaterial use balances its established benefits will only intensify.

Traditionally, emerging technologies have faced risk assessments and the associated public acceptance issues long after products have been commercialized. In this new century, however, public debate about the relative merits of young technologies is starting much earlier in the development process with more substantial impact on the ultimate commercial enterprise. Genetically modified (GM) foods and the Human Genome Project present contrasting cases. GM enthusiasts ignored or dismissed concerns and soon faced a robust and organized activist campaign against GM products. While this has not had a major impact on sales of GM foods in the US, the European market for such products has essentially disappeared and a global debate has ensued over their use in the developing world. In contrast, the founders of the Human Genome Project acknowledged that the scientific discoveries spawned by their research would likely raise societal and ethical concerns and set aside a portion of their research dollars to address them. No significant backlash has developed against HGP. Some of the same critics of GM foods are now turning their attention to nanotechnology. This kind of negative attention at such an early stage in a new technology's development could profoundly impact its ability to gain a foothold in the marketplace.

There is relatively little technical information and proactive public policy concerning the implications of nanotechnology, and this only sustains the arguments of those organizations opposed to developing nanotechnology. While there are some nanotechnology critics that remain concerned about far-term applications such as nanorobots and human-machine hybrids, most focus on the near-term issues of nanoparticle exposure and toxicity. The actions these critics recommend range from calls for new laboratory protocols to protect workers from unknown risks, to a global moratorium on the manufacture and use of engineered nanoparticles until they can be certified as benign to both humans and the environment. [1, 2] Such concerns have received attention from both US and EU policymakers, and several governments are now contemplating an appropriate response. For example, in response to cautionary public statements made by the Prince of Wales, the UK Royal Society was commissioned to carry out an independent study to assess whether nanotechnology is likely to pose new risks not addressed by current regulations. [3, 4] The US House of Representatives passed a bill in 2003 that requires the integration of societal impact studies into nanotechnology research; similar language appears in the pending Senate version of the bill. [5, 6] While these policy deliberations are in an early, exploratory phase, the ultimate outcome may be a new regulatory stance towards nanotechnology aimed at minimizing exposure to the engineered nanoparticles of greatest concern.

Environmental Impact of Engineered Nanomaterials

The question of the environmental impact of engineered nanomaterials is a technical one that scientists and engineers can address experimentally; however, until very recently the topic has received little attention. There are few directly relevant and peer-reviewed studies of engineered-nanoparticle toxicology or environmental impact in the open literature. By 'engineered nanoparticle' we mean nanoparticles with critical dimensions less than 50 nm engineered for a specific function. The modifier 'engineered' is an important one for this work, as it distinguishes nanoparticles created with high perfection and uniformity from the polydisperse, naturally occurring nanoparticles and ultra-fine particles produced in aerosols. The characterization of the environmental impacts of nanomaterials presents many challenges. Most daunting is the extraordinary breadth of nanoscale materials: not only are there many different types, but they can be of many sizes and possess different surface coatings. Moreover, there isn't one 'most important' class of materials on which to focus, nor is there likely to be in the near future. Like polymeric materials, nanomaterials are diverse and will be used in many forms and sizes. In this discussion we limit ourselves to an important class of nanomaterials, namely inorganic, engineered nanoparticles.

A key question to pose from the start is by what routes and in what concentrations will people and wildlife be exposed to engineered nanomaterials? A worker in a fullerene manufacturing plant has a different exposure profile than the bowler whose ball is coated with fullerenes. Clearly, it is essential to characterize the expected concentrations of engineered nanoparticles that may be present in the air, water and soil. A useful way to approach the problem is to consider several likely scenarios for how human populations, both in the present and near future, may be exposed to engineered nanoparticles. Each situation presents different issues for characterizing exposure, and their comparison highlights those scenarios most likely to be relevant for engineered nanomaterials.

Occupational exposure to nanomaterials is growing as demand increases for nanoparticles that add value to consumer products. The number and diversity of companies building plants to manufacture nanoscale materials will only continue to increase as new applications are developed. Mitsubishi's Frontier Carbon Corporation has begun manufacturing multiton quantities of fullerenes and hopes to produce 300 metric tons annually within two years. [7, 8] Frontier envisions fullerenes being incorporated into thousands of applications, including pharmaceuticals, fuel cells, batteries and high-performance coatings. Mass production of a novel material raises different questions about exposure than does small-scale use in the research lab. However, in the US both groups can consult standardized material safety data sheets (MSDS) to assess the hazards of the compounds into which they may come in contact. The MSDS for most nanoparticles are identical to those of the bulk material of equivalent chemical composition. Thus, despite the extensive body of literature demonstrating the novel chemical and physical properties of most nanomaterial, no new formal requirements have been established for safe handling beyond those already in place for the bulk material.

Consumer products containing engineered nanomaterials must also be considered as another potential route of human exposure. Nanoparticles have already been incorporated into personal care products such as sunscreens and cosmetics, where their small size confers benefits such as transparency and increased performance.[9-11] Detailed information on particle size and information is difficult to obtain since the product formulations are considered proprietary by most companies. In 1999, the US Food and Drug Administration considered two petitions regarding the regulation of nanoscopic or "micronized" titania in personal care products. One comment recommended that micronized titania be treated as a "new ingredient with several unresolved safety and efficacy issues." In its final rule, the FDA agreed with the second comment that micronized titania is "not a new material but is a selected distribution of existing material" and that it showed no deleterious effects in animal and human studies. [12] Exposure to micronized titania particles greater than 40 nm through skin absorption has been ruled out in the literature.[13, 14] However, the photocatalytic activity of titania and zinc oxide particles may, through a free radical mechanism, degrade the stability of the organic components of sunscreens, thereby reducing their efficacy.[15-19] Free radical processes can also damage biological tissue, though the risk from sun exposure may be greater.[20, 21] Another question not explicitly addressed by the FDA rule is the relative importance of the physical size of a nanoparticle when considering its risk.

The exposure to nanomaterials of factory workers and consumers of personal care products is a near-term issue worthy of more focused attention by the research community. Equally important is the scarcity of data in the literature regarding exposure mechanisms relevant over the long term, *e.g.*, through the accumulation and transport of nanoparticles in the environment. As nanoparticle manufacture continues to increase, and as more products incorporating them are brought into the marketplace, their concentration in the air, water and soil is likely to rise. Therefore, it becomes important for both the technical and policy communities to understand and address the consequences of that possibility. The technical community can help by evaluating the fate and transport of model systems in these surroundings.

Recommendations

In the absence of reliable scientific data, policymaking may become unduly politicized. Therefore, concrete actions must be taken to ensure that nanotechnology develops responsibly and with strong public support. The most essential component of a responsible policy is a scientific assessment of the impact of nanomaterials on human health and the environment. A body of solid, peer-reviewed data on potential toxicity, bioaccumulation, and fate of nanomaterials in the environment would identify the specific nanomaterials and applications of greatest potential risk so that the risk can be mitigated or avoided altogether. The technical community must also show itself to be responsive to public concerns by engaging in open dialogues with all stakeholders, including nanotechnology's critics, about the best way to ensure that nanotechnology does not have unintended consequences.

References

1. Arnall, A.H., Future technologies, today's choices: Nanotechnology, artificial intelligence and robotics; A technical, political and institutional map of emerging technologies. 2003, Greenpeace Environmental Trust: London, England.

2. ETC Group, No small matter II: The case for a global moratorium. April 2003.

3. Staff, Prince sparks row over nanotechnology, in The Guardian. April 28, 2003: London, England.

4. The Royal Society/Royal Academy of Engineering Study on Nanotechnology, http://www.nanotec.org.uk/index.htm. 2003.

5. U.S. House.108th Congress 1st Session., H.R. 766, Nanotechnology Research and Development Act of 2003. (H.R. 766).

6. U.S. Senate. 108th Congress 1st Session., S. 189, 21st Century Nanotechnology Research and Development Act.

7. Tremblay, J.-F., Mitsubishi chemical aims at breakthrough. Chemical & Engineering News, 2002. 80(49): p. 16-17.

8. Tremblay, J.-F., Fullerenes by the ton. Chemical and Engineering News, 2003. 81(32): p. 13-14.

9. Edwards, M.F. and T. Instone, Particulate products - their manufacture and use. Powder Technology, 2001. 119(1): p. 9-13.

10. Shefer, S. and A. Shefer, Controlled release systems for skin care applications. Journal of Cosmetic Science, 2001. 52(5): p. 350-353.

11. Spiertz, C. and C. Korstanje, A method for assessing the tactile properties of dermatological cream bases. Journal of Dermatological Treatment, 1995. 6(3): p. 155-157.

12. Department of Health and Human Services, Sunscreen drug products for over-the-counter human use; Final monograph. Federal Register. Vol. 64. May 21, 1999: U. S. Government. 1-28 (see p. 6).

13. Lademann, J., et al., Penetration of titanium dioxide microparticles in a sunscreen formulation into the horny layer and the follicular orifice. Skin Pharmacology and Applied Skin Physiology, 1999. 12(5): p. 247-256.

14. Schulz, J., et al., Distribution of sunscreens on skin. Advanced Drug Delivery Reviews, 2002. 54: p. S157-S163.

15. Bahnemann, D.W., et al., Photodestruction of dichloroacetic acid catalyzed by nano-sized TiO_2 particles. Applied Catalysis B-Environmental, 2002. 36(2): p. 161-169.

16. Malato, S., et al., Photocatalysis with solar energy at a pilot-plant scale: an overview. Applied Catalysis B-Environmental, 2002. 37(1): p. 1-15.

17. Ricci, A., et al., TiO_2-promoted mineralization of organic sunscreens in water suspension and sodium dodecyl sulfate micelles. Photochemical & Photobiological Sciences, 2003. 2(5): p. 487-492.

18. Picatonotto, T., et al., Photocatalytic activity of inorganic sunscreens. Journal of Dispersion Science and Technology, 2001. 22(4): p. 381-386.

19. Rossatto, V., et al., Behavior of some rheological modifiers used in cosmetics under photocatalytic conditions. Journal of Dispersion Science and Technology, 2003. 24(2): p. 259-271.

20. Hidaka, H., et al., In vitro photochemical damage to DNA, RNA and their bases by an inorganic sunscreen agent on exposure to UVA and UVB radiation. Journal of Photochemistry and Photobiology a-Chemistry, 1997. 111(1-3): p. 205-213.
21. Dunford, R., et al., Chemical oxidation and DNA damage catalysed by inorganic sunscreen ingredients. Febs Letters, 1997. 418(1-2): p. 87-90.

Environmental Implications: Toxicology and Biological Interactions of Nanomaterials

Editors
John R. Bucher
Vicki L. Colvin

Chapter 5

Environmental Implications: Toxicology and Biointeractions of Nanomaterials

Toxicology and Biological Interactions of Nanomaterials: An Overview

John R. Bucher[1] and Vicki Colvin[2]

[1]National Toxicology Program, National Institute of Environmental Health Sciences, National Institutes of Health, P.O. Box 12233, Research Triangle Park, NC 27709
[2]Center for Biological and Environmental Nanotechnology and Department of Chemistry, Rice University, MS 60, 6100 Main Street, Houston, TX 77005

The papers in this section summarize presentations given in a session on the "toxicology and biological interactions of nanomaterials." The purpose of this session was to review what is currently known of the toxicity of several specific manufactured and naturally occurring nanomaterials, to consider the more general topic of biological interactions of materials on the nanoscale, and to lay the groundwork for development of an effective and focused research program on the collective toxicity of these chemically and physically diverse materials.

28

The symposium session on toxicology and biological interactions of nanomaterials is one of the first of its type. It was designed to attempt to identify potential adverse health effects in the broadest sense that may be associated with the creation, use and or disposal of manufactured nanomaterials.

Currently there are few if any scientists working with a sole focus on the toxicology of manufactured nanomaterials. The closest well established area of investigation is the field of ultrafine particle inhalation toxicology. The growing awareness that adverse cardiovascular and pulmonary health effects of particulate air pollution may be in large part accounted for by the smallest of the combustion particulate byproducts (< 0.1 μm in diameter) is fairly recent (1), and the unexpected finding that nanoparticle size can impact toxicity equally if not more so than chemical composition hints at the complexity of the topic. Indeed, the unusual biological properties of nanomaterials- substances that are neither "chemicals" nor 'particles" as we typically think of them, allows for a collective consideration of a class of substances that differ widely in chemical and physical properties, as well as applications and uses.

It has been proposed that nanoparticles, particles ranging from 1 to 100 nm in any one dimension, were not a significant part of the human environment until combustion of fuels became a central aspect of life (2). This is an interesting hypothesis, although the implication that cells or organisms did not encounter nanomaterials and therefore are ill equipped to deal with them is somewhat implausible. For example, many biological agents such as viruses (3) and lipoproteins that transport fatty acids and lipids fall into the nanometer size range (4), and indeed all organisms have evolved the capability of carrying out the essential processes of life through organization and maintenance of complex functional nanostructures.

However, there are indications in the literature that manufactured nanomaterials or nanoparticles resulting from combustion may interact with biological systems in unpredictable ways. Ultrafine particles deposited in the lung are apparently translocated across the alveolar-capillary barrier through caveolae, rather than taken up by macrophages as is the fate of larger particles, and there is evidence for movement of nanoparticles from the respiratory tract to extrapulmonary organs through neurons (1). The unusual distribution properties of nanoparticles have been exploited by the pharmaceutical industry in developing methods to deliver drugs across the blood-brain barrier (5). In addition, the seemingly unrestricted access of nanoparticles into cells has been exploited through the use of nanocrystals (quantum dots) as fluorescent bioprobes in imaging applications (6). Certain nanoparticles have been observed to preferentially accumulate in particular organelles (7,8). Surface properties seem to play a major role in determining where the materials accumulate within the cell.

Understanding the surface chemistry of nanoparticles may be key to predicting ultimate toxic effects. Surface properties can be changed by coating nanoparticles with different materials, but surface chemistry is also influenced by the size of the particle. Even inert elements such as gold, become more catalytically active when present as nanoparticles (9). It has been shown that inhaled ultrafine polytetrafluoroethylene (PTFE or Teflon®) particles are disproportionately more toxic than larger particles of similar composition but equivalent mass (10). Ultrafine polystyrene particles were reported to induce inflammation in the lung in proportion to their surface area (11). This interaction of surface area and particle composition in provoking biological responses adds an extra dimension of complexity in sorting through the potentially advantageous and adverse events that may result from exposure to these materials.

The speakers chosen for this symposium session were asked to address both recognized toxicities as well as apparently benign interactions of nanoparticles with biological systems. Dr. Paul Alivisatos of the University of California, Berkeley, and Lawrence Berkeley National Laboratory led off the technical portion of the program by introducing quantum dots; outlining his laboratory's bioimaging studies with these materials, and describing some applications in determining the metastatic potential of cells in culture. Dr. Bernard Erlanger of Columbia University discussed a remarkable series of studies wherein antibodies have been raised to chemically-modified fullerenes and were shown to cross react with single walled carbon nanotubes (12, and see commentary 13). These studies have applicability to the question of the potential of manufactured nanoparticles to produce autoimmune disease and other immunological responses. Dr. David Allen of Texas Tech then reviewed studies from his laboratory examining the interactions of nanoparticles, under development as drug carriers, on the function of the blood-brain barrier.

The second half of the symposium session dealt with examinations of toxicity's resulting from exposures to nanoscale materials. Dr. Gunter Oberdörster, University of Rochester, reviewed a series of studies his laboratory has carried out looking at the toxicity, fate and transport of various inhaled ultrafine particles. These studies form a large part of the existing knowledge base on particle fate and transport outlined above. Finally, Drs Chiu-wing Lam, Johnson Space Center, and David Warheit, DuPont Haskell Laboratory, described a set of independently performed studies that both examined the consequences of intratracheal instillation to rats of different preparations of single wall carbon nanotubes. Although the pulmonary lesions produced in these studies were qualitatively similar, the investigators held different opinions concerning the interpretations of the studies, and on the implications of these findings for human health following inhalation of these materials.

In our estimation, this symposium session established that there is merit in considering as a collective concept the toxicity of nanomaterials of a wide variety of chemical and physical properties. It is our hope that recognition of the unique characteristics of nanoparticles will enable us to design studies that will establish critical toxicologic principles. This information could aid in the design of toxicologically benign materials and provide the science base to support safe implementation and societal acceptance of a promising technology.

This symposium was reviewed in articles published in Science (*14*), and Chemical & Engineering News (*15*).

References

1. Oberdörster, G.; Utell, M.J. Ultrafine particles in the urban air: To the respiratory tract- and beyond? *Environ. Health Perspec.* **2002**, *110*, A440-1.

2. Howard, C. V. Nanoparticles and Toxicity. ETC Group Occasional Paper Series: Winnipeg, Canada 2003, vol *7*, No. 1. Annex.

3. Masciangioli, T.; Zhang, W-X. Environmental technologies at the nanoscale. *Environ. Sci. Technol.* **2003**, *Mar. 1*, 102A- 108A.

4.. Badiou, S.; Merle De Boever, C.; Dupay, A.M.; Balliat, V.; Cristol, J.P.; Reynes, J. Decrease in LDL size in HIV positive adults before and after lopinavir/ritonavir-containing regimen: an index of atherogenicity? *Atherosclerosis* **2003**, *168*, 107-113.

5. Lockman, P.R.; Mumper, R.J.; Kahn, M.A.; Allen, D.D. Nanoparticle technology for drug delivery across the blood-brain barrier. *Drug Develop. Ind. Pharmacy* **2002**, *28*, 1-12.

6. Seydel, C. Quantum dots get wet. Science **2003**, *300*, 80-81.

7. Foley, S.; Crowley, C.; Smaihi, M.; Bonfils, C.; Erlanger, B. F.; Seta, P.; Larroque, C. Cellular localisation of a water-soluble fullerene derivative. *Biochem. Biophys. Res. Commun.* **2002**, *294*, 117-119.

8. Savic, R.; Luo, L.; Eisenberg, A.; Maysinger, D. Micellular nanocontainers distribute to defined cytoplasmic organelles. *Science*, **2003**, *300*, 615-618.

9. Haruta, M. When gold is not noble: catalysis by nanoparticles. *Chem. Rec.* **2003**, *3*, 75-87.

10. Johnston, C.J.; Finkelstein, J.N.; Mercer, P.; Corson, N..; Gelein,, R.; Oberdörster, G. Pulmonary effects induced by ultrafine PTFE particles. *Toxicol. Appl. Pharmacol.* **2000**, *168*, 208-215.

11. Brown, D.M.; Wilson, M.R.; MacNee, W.; Stone, V.; Donaldson, K. Size-dependent proinflammatory effects of ultrafine polystyrene particles: a role for surface area and oxidative stress in the enhanced activity of ultrafines. *Toxicol. Appl. Pharmacol.* **2001**, *175*, 191-199.

12. Erlanger, B. F; Bi-Xing, C.; Zhu, M.; Brus, L. Binding of an anti-fullerene IgG monoclonal antibody to single wall carbon nanotubes. *Nano Lett.* **2001**, *1*, 465-467.

13. Izhaky, D.; Pecht, I. What else can the immune system recognize? *Proc. Natl. Acad. Sci. USA* **1998**, *95* 11509-11510.

14. Service, R. F. Nanomaterials show signs of toxicity. *Science* **2003**, *300*, 243.

15. Dagani, R. Nanomaterials: Safe or unsafe? *Chem. Engineer. News* **2003**, *April 23*, 30-33.

Chapter 6

A Role for Immunology in Nanotechnology

Bernard F. Erlanger

Department of Microbiology, Columbia University, New York, NY 10032

Until 1985, only two allotropic forms of elemental carbon were known: graphite and diamond. In 1985, a third form was discovered constructed of 60 carbon atoms (i.e. C_{60}) and named Buckminsterfullerene because of its geodesic character (*1*). The preparation of fullerenes in workable quantities led first to applications in chemical and engineering processes and subsequently to the suggested use of fullerenes for biological and medical applications (*2-4*) and as templates for the design of experimental pharmaceutical agents, including those with anti-viral (*5-9*), antioxidant (*8-12*), chemotactic (*13*), and neuroprotective (*14*) activities.

We demonstrated that the mouse immune repertoire is diverse enough to recognize and produce antibodies specific for C60-fullerenes (*15*). We succeeded in isolating several monoclonal anti-C $_{60}$ antibodies. The monoclonal antibody under study (1-10F-A8) is an IgG1, kappa and was prepared by standard procedures which included immunization of mice with a fullerene carboxylic acid derivative covalently linked to a protein (bovine thyroglobulin) by competitive inhibition in an ELISA format, in which a fullerene-rabbit serum albumin (RSA) conjugate was used as the target. An apparent binding constant of 22 nM was determined. The sequences of the light and heavy chains were determined, and the three-dimensional structure of the Fab fragment was solved and refined by x-ray crystallographic techniques to a resolution of 2.25 A. Finally, we identified and modeled the probable binding site for C 60 fullerene and described the interatomic interactions that stabilize the antibody-fullerene complex.

Identification and Modeling of C $_{60}$ Binding to the Fab' Fragment of Monoclonal Antibody 1-10F-A8 by X-ray Crystallization

Identification of the C $_{60}$ binding site in the anti-fullerene antibody was accomplished by X-ray crystallization using accessible surface area calculation and docking/energy minimization procedures. Accessible surface area calculation with a 1.7 A radius probe identified a spherical-shaped cavity in the interface of the two sequence-variable peptide chains, VH and VL, that make up the binding site of an antibody molecule. The two chains contain within them amino acids that actually contact the antigen, defined as the Complementary Determining Regions (CDRs). There are also the "framework" regions that define the combining site but do not make contact with the fullerene. The V L CDR amino acid residues are Tyr-36, Gln-89, Phe-96 and the V L "framework" residue Phe-98. Included also are the side chains of VH CDR residue Asn-35 and VH framework residues Val-37, Trp-47, and Trp-106. In addition the main chain atoms of the V H residues Ala-97, Thr-98 and Ser-99 (in CDR H3) also define the cavity. The cavity contains two solvent water molecules that form a solvent bridge between the side chains of V L Gln-89 and V H Asn-35.

Accessibility to the fullerene binding site is via a 4.7 A wide channel fashioned by the CDR L3 and the short CDR H3 hypervariable loop. The cavity identified by the V L-V H interface is approximately 7 A in diameter, too small for the C_{60} fullerene of 10 A diameter, incorporating the van der Waals radii of the carbon atoms. Nonetheless, manual docking of a model C_{60} into the cavity and using an energy minimization protocol results in relaxation of the surrounding Fab side chains and 11 degree rotation of the VL and VH domains. The binding process therefore, includes an induced fit mechanism. The antibody surface area buried by the V L and V H interface is relatively small, 1,100 A^2, consistent with other anti-hapten antibodies that undergo large relative V L -V H displacements upon hapten binding.

In addition to confirming that the fullerene binding site was the spherical cavity identified by accessible surface area calculation, the modeling and energy minimization of the Fab-C_{60} complex suggests several pi-bond stacking interactions between the fullerene and the antibody. Aromatic side chains of VHTrp-47, VL Tyr-91, and VL Phe-96 as well as side chains for VH Asn-35 and VL Gln-89 lie parallel in the C_{60} molecule. In addition, the potential for a weak hydrogen bond from the VL Tyr-36 hydroxyl to the fullerene (3.15 A) was also noted. Stacking interactions to C_{60} have been previously described in the crystal structure of C_{60} solvated by benzene. The solvent benzene was noted to lie over the electron-rich C_{60} pentagon-pentagon bonds, clearly establishing the nature of the stacking interaction. Pi-system stacking interactions are well established in x-ray crystal structures of antibody-antigen complexes. For example, in the antibody-antigen complex of the anti-lysozyme antibody D44, the stacking interaction involved an aromatic side chain, tryptophan, from the antibody and an arginine from the antigen. It is not surprising, therefore, that the anti-fullerene antibody uses similar stacking interactions for the binding of fullerene.

It should be noted that the stacking and hydrogen bond interactions to the fullerene would be weak interactions.

Binding of the Anti-Fullerene Antibody to Single Wall Carbon Nanotubes

Single wall carbon nanotubes (SWNTs) are a remarkable new class of nanometer diameter metallic and semiconducting wires that carry current as pi electrons propagating on their graphitic surface. They are physically robust, exhibit great tensile strength, do not oxidize or have surface states under ambient conditions, and show high conductivity. They are easily grown in lengths of tens of microns and can be precisely positioned and manipulated when attached to AFM tips. Their remarkable electrical properties suggest their potential for use in future electronic components.

The immunochemical binding interreaction between SWNTs and the C60 antibody was first demonstrated by a competitive ELISA in which a colloidal suspension of carbon nanotubes was shown to competitively inhibit the binding of anti-fullerene antibody to a fullerene conjugate of rabbit serum albumin. Competition was seen at very high dilutions of the SWNT colloidal suspension. A quantitative measure of binding coefficient will require detailed study as a function of available SWNT surface area, structural type, and extent of SWNT aggregation into ropes.

Binding was confirmed directly by atomic force microscopy. SWNT ropes on mica were initially imaged, the surface was then exposed to antibody solution, and finally, the same SWNT was imaged in air. This sequence distinguishes any preexisting surface particles from bound antibodies. Three drops of aqueous SWNT suspension (0.064 mg/mL in Triton 100X surfactant, from Tubes@Rice) were spun into freshly cleaved mica surface. The sample was imaged by tapping mode AFM in air. Many anti-fullerene antibody molecules could be seen adsorbed to the surface of nanotubes (*16*).

Thus the monoclonal anti-C 60-specific antibody could be visualized binding to carbon nanotubes, presumably because the latter have a curved, hydrophobic pi-electron-rich surface, analogous to C60, itself. The hydrophobic binding site of the antibody is sufficiently flexible to recognize both. Thus our findings bridged 2 disparate disciplines, electrical nanotechnology and monoclonal immunology. We were encouraged, therefore, to explore the possibility of utilizing this combination in biological systems, in particular with the idea that antibody–coated SWNTs can be used advantageously as probes of cell or membrane function. A SWNT rope has a diameter of about 10nm, far smaller than present metal or glass capillary intracellular probes. They should be capable of insertion into and withdrawal from specific regions of cells, with minimal disturbance of cell or membrane function.

We first showed that the monoclonal anti-fullerene antibody could be covalently "decorated" with a fluorescent Ca^{++} probe (Molecular Probes, Eugene OR) without disturbing its specificity for fullerenes and nanotubes. Our future plans envision insertion of the SWNT- antibody-calcium probe into living cells. It should be possible for the probe molecule(s) to be optically excited or electrically addressed via the conducting SWNT wire. Unlike most semiconductors and metals, SWNTs do not form insulating surface oxides at room temperature, i.e. there should be electrical contact with the antibody. Indeed, recent experiments by others have demonstrated that nanotube electrical properties change with reversible adsorption of molecular species.

Acknowledgment

(Funding was from CSixty , Inc., Toronto Canada and the National Institutes of Health).

References

1.   Kroto, H.W. et al. *Nature* **1985**, *318*, 162-163 .
2.   Jensen, A.W. et al. *Bioorg. Med. Chem.* **1996**, *4*, 767-779.
3.   Da Ros, T.& Krato, M. *Chem. Commun.* **1999**, 663-669.
4.   Wilson, S.R. *The Fullerene Handbook* Wiley, N.Y. in press
5.   Sijbesma, R. et al. *J. Amer. Chem Soc.***1993**, *115*, 6510-6512.
6.   Schinazi, R.F. et al. *Antimicrob. Agents Chemother.* **1993**, *37*, 1707-1710.
7.   Friedman, S.H. et al. *J. Med. Chem.* **1998**, *41*, 2424-2429.
8.   Chang, L.Y. et al. *Chem. Commun.* **1995**, 1283-1284.
9.   Lu, L.H. et al. *Br. J. Pharmacol.* **1998**, *123*,1097-1102.
10.   Straface, E. et al. *FEBS Lett.* **1999**, *454*, 335-340.
11.   Lin, A.M. et al. *J. Neuro. Chem.* **1999**, *72*:,1634-1640.
12.   Chueh, S.C. et al. *Transplant Proc.* **1999**, *31*, 4558-4562.
13.   Tonolo, C. et al. *J. Med Chem.* **1994**, *37*, 4558-4562.
14.   Dugan. L.L. et al. *Proc. Natl Acad. Sci. USA* **1997**, *94,* 9434-9439.
15.   Chen, B.-X. et al. *Proc. Natl Acad. Sci. USA* **1998**, *95*, 10809-10813.
16.   Erlanger, B.F. et al. *Nano Letters* **2001**, *1,* 465-467.

Chapter 7

Effects and Fate of Inhaled Ultrafine Particles

Günter Oberdörster

Department of Environmental Medicine, University of Rochester,
575 Elmwood Avenue, Rochester, NY 14642
(Gunter-Oberddorster@urmc.rochester.edu)

Concentration of inhaled particulate matter (PM) varies by orders of magnitude from ng/m^3 to mg/m^3, depending on material, particle size, and proximity to sources. Particle number concentrations, expressed as $particles/cm^3$, can also span several orders of magnitude, with ultrafine particles (<100 nm) accounting by far for the highest number yet lowest mass concentrations. Deposition in extrapulmonary and pulmonary airways is mainly governed by impaction, settling, diffusion and interception. The nose is a very efficient filter for large (>2 μm, impaction) and smaller ultrafine (<5 nm, diffusion) particles; in the respiratory tract overall there is minimal deposition at 0.2 – 0.5 μm (settling and diffusional displacement minimal). While chemical composition does not influence deposition of inhaled particles—assuming no hygroscopicity—disposition of deposited particles is highly dependent on their chemical characteristics. Since PM is a chemically complex mixture, different mechanisms including dissolution, leaching, chemical binding and mechanical transport are involved in particle clearance. For solid poorly soluble particles, classical clearance mechanisms are size dependent and pathways include mucociliary escalator in nasal and tracheobronchial airways to the GI-tract, alveolar macrophage-mediated clearance in the alveolar region to the mucociliary escalator, interstitial translocation and lymphatic uptake to regional lymph nodes. More recently, two additional pathways to extrapulmonary organs have been described, specifically for poorly soluble ultrafine particles: translocation to blood circulation; and transport to ganglia and structures of the CNS along axons of sensory nerves located in nasal and tracheobronchial epithelia. This could provide a plausible mechanism for adverse cardiovascular and CNS health effects.

Introduction

Several epidemiological studies have found associations of ambient ultrafine particles (UFP) with adverse respiratory and cardiovascular effects resulting in morbidity and mortality in susceptible parts of the population (1–6) whereas other epidemiological studies have not seen such associations (7, 8). Controlled clinical studies evaluated deposition and effects of laboratory-generated UFP. High deposition efficiencies in the total respiratory tract were found as well as effects on the cardiovascular system (9–13). Studies in animals using laboratory-generated surrogate UFP or ambient UFP showed that UFP consistently induced mild yet significant pulmonary inflammatory responses as well as effects in extrapulmonary organs. Animal inhalation studies included the use of different susceptibility models in rodents, with analysis of lung lavage parameters and lung histopathology, effects on the blood coagulation cascade and translocation studies to extrapulmonary tissues (14–21). In vitro studies by a number of different investigators using different cell systems showed to varying degrees oxidative stress–related cellular responses after dosing with laboratory-generated or filter collected ambient ultrafine particles (22–25). Although effects of UFP were consistently observed it appears that more data are needed for their toxicological characterization specifically with regard to developing occupational or environmental standards.

However, despite uncertainties regarding ambient UFP effects, there are a number of characteristics which are unique for UFP. These relate to their behavior in the air and in the respiratory tract, giving them a distinctly different potential to cause adverse health effects. The following short treatise provides an overview of these UFP-specific characteristics.

Respiratory Tract Deposition

The main mechanism of deposition of inhaled UFP in the respiratory tract is by diffusion due to displacement when colliding with air molecules. Other deposition mechanisms, inertial impaction, gravitational settling, and interception, do not contribute to UFP deposition, and electrostatic precipitation only in cases where UFP will carry significant electric charges. The predictive ICRP model (26) gives the fractional deposition of inhaled particles in the nasopharyngeal, tracheobronchial and alveolar region of the human respiratory tract under conditions of nose-breathing during rest. For ultrafine particles (1–100 nm), in each of the three regions of the respiratory tract different amounts of a given size of these particles are deposited. For example, 90% of inhaled 1 nm particles are deposited in the nasopharyngeal compartment whereas it is only ~10% in the tracheobronchial region and essentially none in the alveolar region. On the other hand, 5 nm particles show about equal deposition of ~30% of the inhaled particles in all three regions; 20 nm particles have the highest deposition efficiency in the alveolar region (~50%) whereas in tracheobronchial and nasopharyngeal regions this particle size deposits with ~15% efficiency. These different deposition efficiencies should also have consequences for potential effects induced by inhaled UFP of different sizes as well as for their disposition to extrapulmonary organs, as will be discussed later.

The ICRP model predictions are based on experimental evidence of numerous well-conducted studies in humans for particles above 0.1 μm for all three regions of the respiratory tract (for review see reference 27). With respect to ultrafine particles, experimental data for nasal deposition in humans have been published (28, 29), which served as a basis for the ICRP model. There are also UFP deposition data in humans for tracheobronchial and the alveolar region of the respiratory tract (12, 30), and there are human deposition data for inhaled UFP for total respiratory tract deposition (10, 31–33).

The information about the fraction of inhaled particles depositing in different regions of the respiratory tract could be interpreted as showing that a high deposition fraction in one region of the respiratory tract also implies a high dose to individual cells in that region. However, large differences in epithelial surface area between the different regions need to be considered. For example, inhaled 20 nm particles have the highest deposition efficiency in the alveolar region of the lung. When modeling the deposition of such particles along the individual generations of the lower respiratory tract, one can see that, indeed, most of the mass of these particles is depositing beyond generation 16 of the tracheobronchial region, i.e., in the alveolar region (Fig. 1a). However, considering that the epithelial surface areas in tracheobronchial and alveolar regions have vastly different sizes, one needs to take into account the deposited dose normalized per unit surface area. Figure 1b shows that in that case the

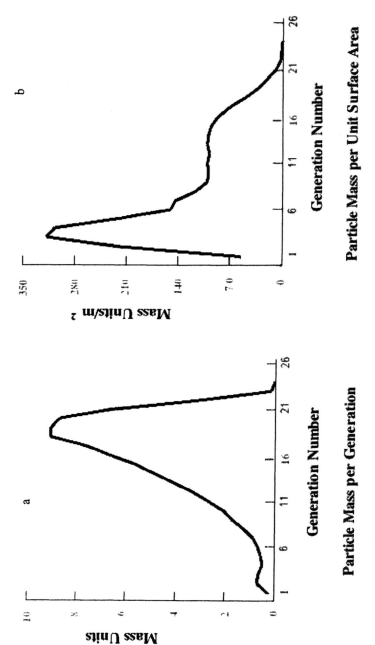

Figure 1: Deposition of inhaled 20 nm particles in human tracheo-bronchial and alveolar region showing deposited mass per airway generation and per unit surface area of each generation. Generation 1 = trachea; Generation 16 = terminal bronchioles. *(Predicted by MPPD model, CIIT/RIVM, 2002).*

upper generations of the tracheobronchial region, in fact, receive the highest doses per unit surface area. An additional factor that needs to be considered is that during deposition hot spots of deposition occur at bifurcational junctions (34, 35). These hot spots comprising about 100 cells or areas of 100 x 100 μm on carinal ridges, can increase focal concentrations by 1-2 orders of magnitude.

Another example may serve to illustrate further differences between regional and surface area deposition, while at the same time comparing the deposition of inhaled poly-disperse ultrafine 20 nm and fine 250 nm particles in all three regions of the human respiratory tract (Figure 2). Assumed is an inhaled concentration of 100 μg/m^3 over a 6-hr. exposure period. The geometric standard deviation of the particle size distribution is 1.7. Deposited amounts, or dose, per region and per unit surface area in all three regions of the respiratory tract are shown, as predicted by the MPPD model (36). As can be seen, the ultrafine particle deposition on a mass basis is more than twice in each of the three regions compared to the 250 nm fine particle. The dose deposited per region increases from the nasal to the tracheobronchial to the alveolar region for both particle types. When expressing the deposited dose per unit surface area, the picture is reversed. While the more than 2-fold greater deposition of UFP is unchanged, the highest surface area dose is received by the nasal followed by the tracheobronchial region, and the least is deposited in the alveolar region (Figure 2). Expressed in terms of number of particles per unit surface area, it is about 5,000 times higher for the 20 nm UFP compared to 250 nm fine particles. This may have significant implications for the likelihood of ultrafine particles to be translocated to extrapulmonary sites as will be discussed later.

Respiratory Tract Clearance

A number of defense mechanisms exists throughout the respiratory tract aimed at keeping the mucosal surfaces free from cell debris and foreign materials, including particles deposited by inhalation. Several reviews describe the well-known classic clearance mechanisms and pathways for deposited particles, (e.g., 27, 37, so the following paragraphs will only briefly mention those mechanisms and point out specific differences that exist with respect to ultrafine particles. A major focus in this article is on an additional not generally recognized mechanism which appears to be very specific for ultrafine particles and which involves their translocation from the respiratory tract to the Central Nervous System (CNS).

The clearance of deposited particles in the respiratory tract is basically due to two processes: physical translocation of particles and chemical dissolution or leaching. The latter mechanism is directed at biosoluble particles or components

42

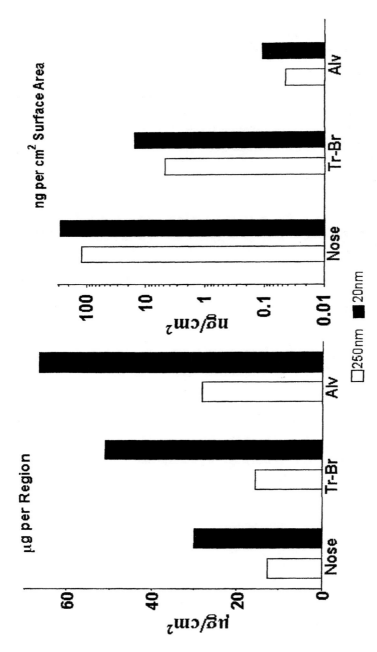

Figure 2: Deposited doses in human naso-pharyngeal, tracheobronchial and alveolar region of aerosol with 20 and 250 nm (CMD) particles (GSD = 1.7) inhaled over 6 hrs. at 100 μg/m³ *(predicted by MPPD model)*.

of particles that are either lipid soluble or soluble in intracellular and extracellular fluids. Solutes and soluble components can then undergo adsorption and diffusion or binding to proteins and other subcellular structures and maybe eventually cleared into blood and lymphatic circulation. This mechanism of clearance for biosoluble materials can happen at any location within the three regions of the respiratory tract. In contrast, a number of diverse processes involving physical translocation of inhaled particles exists which are different in the three regions of the respiratory tract. Figure 3 summarizes these clearance processes for solid particles. As will be pointed out, some of them show significant particle size–dependent differences, making them either uniquely effective for a given particle size or very inefficient for another size.

Nasal mucosa and tracheobronchial region are supplied with an effective clearance mechanism consisting of ciliated cells forming a mucociliary escalator to move a mucus blanket towards the pharynx (posterior nasal region and tracheobronchial region). This constitutes a very fast clearance for solid particles which removes most of them within 24 hours in the tracheobronchial region. It operates most likely also for ultrafine particles (21), although studies by Schürch et al., (38) et al. show that surface tension lowering forces of a thin surfactant layer in the bronchial tree act on particles to submerge them into the mucus and sol phase of the airway fluid which may result in a prolonged retention in this region (39). From the oropharynx, participles are then swallowed into the GI tract thus being eliminated from the respiratory tract. A more detailed review of these nasal mechanisms is provided by Kreyling and Scheuch (37).

The most prevalent mechanism for solid particle clearance in the alveolar region is mediated by alveolar macrophages, through phagocytosis of deposited particles. The success of macrophage-particle encounter appears to be facilitated by chemotactic attraction of alveolar macrophages to the site of particle deposition (40). The chemotactic signal is most likely C5a, derived from activation of the complement cascade from serum proteins present on the alveolar surface (41, 42). This is followed by gradual movement of the macrophages with internalized particles towards the mucociliary escalator. The retention halftime of solid particles in the alveolar region based on this clearance mechanism is about 70 days in rodents and up to 700 days in humans. By 6–12 hours after deposition in the alveolar region, essentially all of the particles are phagocytized by alveolar macrophages, provided, they are of the right size. It appears that there are significant particle size–dependent differences in a cascade of events leading to effective alveolar macrophage phagocytosis.

Figure 4 displays results of several studies in which rats had been exposed to different sized particles (for the 3 and 10 μm particles intratracheal instillation of 40 and 10 μg polystyrene beads were used). Twenty-four hours later, ~80% of 0.5, 3 and 10 μm particles could be retrieved by lung lavage with

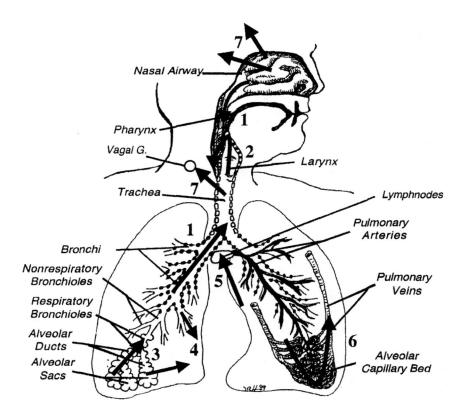

Figure 3: Respiratory tract particle clearance pathways. (1) Mucociliary escalator; (2) GI tract; (3) AM-mediated clearance; (4) Interstitium (*via* epithelium); (5) Lymphat. circulation; (6) Blood circulation; (7) Sensory neurons (olfactory, trigeminus, t-bronchial).

45

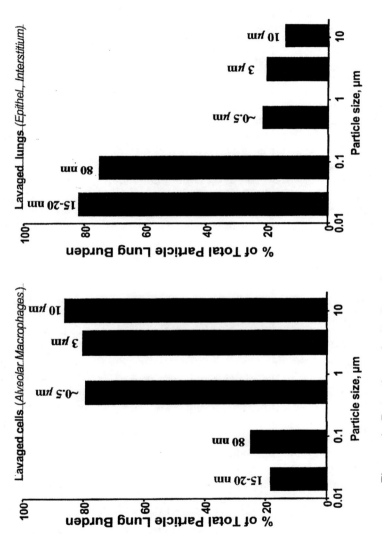

Figure 4: Retention of ultrafine, fine and coarse particles in alveolar macrophages of rats determined 24 hrs. post-exposure by exhaustive lung lavage.

the macrophages, whereas only ~20% of ultrafine 15–20 nm and 80 nm particles could be lavaged with the macrophages. Conversely, ~80% of the ultrafine particles were retained in the lavaged lung after exhaustive lavage, and ~20% of the larger particles > 0.5 µm remained in the lavaged lung.

These results imply that UFP—inhaled and deposited as singlets in the alveolar space—are not efficiently phagocytized by alveolar macrophages which could be either due to an inability of macrophages to phagocytize these small particles or due to a lack of the deposited singlet UFP to generate a chemotactic signal at the site of their deposition. Studies seemingly supporting the first hypothesis, indeed, show that an optimal particle size for phagocytosis by alveolar and other macrophages is between 1–3 µm, and that beyond these sizes phagocytosis rates become progressively slower (43–45). However, in vitro dosing of alveolar macrophages with ultrafine particles indicates that they are phagocytized by macrophages and activate these cells (23, 46; 24; 22). Since in vitro studies do not mimic realistic in vivo conditions, the second hypothesis is more likely to explain the result in Figure 4, that the macrophages do not "sense" the deposited UFP since they did not generate a chemotactic signal or only a very weak one. This suggestion is also supported by the low deposited dose per unit alveolar surface area even for those UFP (20 nm) that have the highest deposition efficiency there (Fig. 1b). Still, experimental proof for this hypothesis is needed to explain the results of Figure 4 that UFP deposited by inhalation in the alveolar region are not efficiently phagocytized by alveolar macrophages.

Because of the inefficiency of alveolar macrophage uptake of UFP one might expect that UFP interact instead with epithelial cells. Indeed, results from several studies show that UFP deposited in the respiratory tract readily gain access to epithelial and interstitial sites (37). We showed in studies with ultrafine PTFE fumes that shortly after a 15-min. exposure the fluorine-containing particles could be found in interstitial and submucosal sites of the conducting airways as well as in the interstitium of the periphery close to the pleura (20). Results of other studies with ultrafine and fine TiO_2 particles—ranging in size from 12 nm to 250 nm—are consistent with a strongly particle size dependent translocation to epithelial and interstitial sites: 24 hours after intratracheal instillation of 500 mg into rats, more than 50% of the 12 nm sized TiO_2 had translocated, whereas it was only 4% and less for 220 nm and 250 nm TiO_2 (47). Of course, the smaller particle sizes of 12 and 20 nm vs. 220 and 250 nm also means that the administered particle number was more than 3 orders of magnitude higher for the ultrafines, a factor that seems to be an important determinant for particle translocation across the alveolar epithelium, as are the delivered total dose and the dose rate (48). In general, interstitial translocation of fine particles across the alveolar epithelium is more prominent in larger species (dogs, non-human primates) than in rodents (37, 49), and one may, therefore, assume that the high UFP translocation observed in rats are likely to occur in humans as well.

Interstitial translocation constitutes another translocation pathway for those particles which are not phagocytized by alveolar macrophages, either due to their small size—which is the case for UFP—or due to an overloading of the alveolar macrophage capacity to phagocytize particles. A state of particle overload has been induced in a number of chronic rat inhalation studies with very high particle concentrations leading to increased translocation also of fine particles (50). Once in the interstitium, translocation to regional lymph nodes either as free particles or after phagocytosis by interstitial macrophages can occur.

Some particles after accumulation in lymph nodes will also translocate further into post-nodal lymph and entering the blood circulation. This mechanism was demonstrated for fibrous particles in dogs: Intrabronchially instilled amosite fibers were found in post-nodal lymph samples of the right thoracic duct collected in the neck area before entering the venous circulation (51). There was an obvious size limitation in that only the shorter and thinner fibers (< 500 nm diameter) appeared in the post-nodal lymph. This lympho-hematogeneous pathway appears to exist also in non-human primates where it was shown that fine crystalline silica particles translocated to the livers of exposed monkeys following chronic exposure to high concentrations (52. 53). Thus, the clearance pathway from local lymph nodes to the blood circulation is not restricted to UFP but occurs also for larger particles, although rates are likely different. The initial step, however, involving transcytosis across the alveolar epithelium into the pulmonary interstitium, seems to occur with larger particles only in high lung load situations when the phagocytic capacity of alveolar macrophages is overwhelmed and the particles are present in the alveoli as free particles for an extended period of time.

Once the particles have reached pulmonary interstitial sites, uptake into the blood circulation in addition to lymphatic pathways can occur, a pathway that again is dependent on particle size, favoring ultrafine particles. Several recent studies in rodents and humans indicate that rapid translocation of inhaled UFP into the blood circulation occurs.

Nemmar et al. (14) reported findings in humans that inhalation of ^{99m}Tc-labelled ultrafine carbon particles (Technegas®) resulted in the rapid appearance of the label in the blood circulation shortly after exposure and also in the liver. They suggested that this at least partly indicated translocation of inhaled UFP into the blood circulation. In contrast, other studies in humans with ^{99m}Tc-labeled carbon particles (33 nm) by Brown et al. (23) did not confirm such uptake into the liver, and the authors cautioned that the findings by Nemmar et al. (14) were likely due to soluble pertechnetate rather than labeled UFP. Our inhalation studies in rats have shown that ultrafine elemental ^{13}C particles (CMD ~30 nm) had accumulated to a large degree in the liver of rats by 24 hours after exposure, indicating efficient translocation into the blood circulation (54). Suggested pathways into the blood could be across the alveolar epithelium as well as across intestinal epithelium from particles swallowed into the GI tract. On the other hand, using a method of intratracheal inhalation of ultrafine ^{192}Ir particles in rats, Kreyling et al. (37) found only minimal translocation (<1%) from the lung to extrapulmonary organs, although they reported 10-fold greater translocation of the smaller (15 nm) compared to the larger (80 nm) UFP.

The conclusion from these studies is that translocation of inhaled UFP from the respiratory tract into the blood circulation can occur, although the efficiency of such translocation may well depend on the chemical and surface characteristics of the UFP. Additional studies are needed to determine particle size and surface chemistry dependent translocation rates across the alveolar epithelial/endothelial cell barrier. With respect to the mechanism of this translocation pathway, it has been suggested that caveolae existing in alveolar epithelial and endothelial cell membranes support this transcytosis process (55). During inspiratory expansion and expiratory contraction of the alveolar walls these caveolae with openings around 40 nm disappear and reappear, forming vesicles which are thought to function as transport pathways across the cells for macromolecules (56). Caveolin proteins associated with these caveolae may also be involved in this translocation process.

Finally, a translocation pathway in the respiratory tract involving neuronal axons should be discussed here. This pathway may be specific for UFP and was already discovered more than 60 years ago, yet it has received only little or no attention by inhalation toxicologists. It is depicted in Figure 3 for the nasal and tracheobronchial regions, comprising sensory nerve endings of the olfactory and the trigeminus nerves and of an intricate network of sensory fibers in the tracheobronchial region. Studies in the early 1940's (57, 58) showed that 30 nm poliovirus particles intranasally instilled into chimpanzees translocated along the axon of olfactory nerves into the olfactory bulb. DeLorenzo (59) in later studies related to the physiology and anatomy of olfaction provided excellent EM photographs of colloidal 50 nm gold particles translocating in the axon of olfactory nerves into the olfactory bulb of squirrel monkeys following intranasal instillations. DeLorenzo could manifest the translocation of these electron dense gold particles from their uptake into the olfactory rods, their retrograde movement along the dendrites, anterograde movement along the olfactory nerve axons to the olfactory bulb, even crossing synapses to the second layer of mitral cells in the bulb, and finally the localization of the 50 nm particles in mitochondria. Both groups of investigators also determined the transport velocity of the 30 nm virus and 50 nm gold particles at 2.4 and 2.5 mm/hour, at which speed the particles appeared within 30-60 mins. after nasal inoculation in the olfactory bulb. This agrees very well with neuronal transport velocities of solid particles in axons of other neurons after direct microinjection (60).

Hunter and Dey (62) and Hunter et al. (63) documented in subsequent studies with intranasally and intratracheally instilled rhodamine labeled microspheres neuronal uptake and translocation *via* trigeminus nerves in the nasal region and sensory nerves in the tracheobronchial region. The particles were transported to the trigeminal ganglion at the internal base of the skull and to the ganglion nodosum in the neck area (with connections to the vagal nerve), respectively. These findings are based on the now well-established method of using 20-200 nm microspheres, which are readily transported in axons and dendrites as tracers to visualize neurons. This axonal/dendritic transport of microspheres was first described by Katz et al. (64) after microinjection into brain tissue of rats and cats. Size and surface properties of the microspheres seem to influence the neuronal translocation process, the optimal size possibly being ~30 nm UFP (*Katz, personal communication*).

In a most recent study we have found that inhalation of ultrafine elemental ^{13}C particles (35 nm) resulted in their significant accumulation over the 7-day post-exposure period in the olfactory bulb of rats (Fig. 5). On two of the post-exposure days there was also a significant increase of ^{13}C in cerebrum and cerebellum, however, the olfactory bulb showed the most consistent significant increases (65). Preliminary results of another inhalation study with ultrafine (30 nm) manganese oxide particles in rats again showed significant increases of Mn in the olfactory bulb as well as striatum and frontal cortex. These studies are consistent with the olfactory nerve translocation route demonstrating that this pathway is also operating for inhaled UFP deposited in the nasal region of the rat.

Significance of Translocation from the Respiratory Tract

The preceding sections summarized data demonstrating that inhaled UFP of different sizes can target all three regions of the respiratory tract. Once deposited, these particles—in contrast to larger-sized particles—appear to translocate readily to extrapulmonary sites and reach other target organs by different transfer routes and mechanisms. One involves transcytosis across epithelia of the respiratory tract into the interstitium and access to the blood circulation directly or via lymphatics resulting in distribution of UFP throughout the body. The other is uptake by sensory nerve endings embedded in airway epithelia, followed by neuronal translocation to ganglionic and CNS structures.

Translocation of inhaled UFP into the blood circulation could affect endothelial cell function as suggested by results of Nemmar et al. (15; 66). These authors saw in an experimental thrombosis model in hamsters that i.v. injected and intratracheally instilled positively, but not negatively, charged 60 nm ultrafine polystyrene particles increased thrombus formation in a peripheral

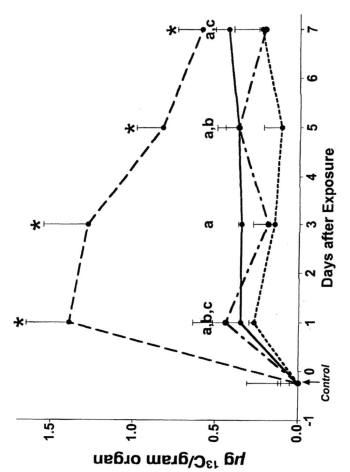

Figure 5: Time course of ^{13} tissue concentrations in lung, olfactory bulb, cerebrum and cerebellum of rats following a 6-hr. inhalation exposure to ultrafine (36 nm CMD) elemental ^{13}C particles (n = 3 rats per timepoint). -- lung; —— olfactory; — · — cerebrum; .... cerebellum * and a, b, c indicate values significantly greater than controls, p < 0.05 (ANOVA) (* = for lung; a = for olfactory bulb; b = for cerebellum; c = for cerebrum).

vein. Larger 400 nm positively charged particles did not increase thrombus formation when intratracheally instilled at a dose which induced pulmonary inflammation equivalent to that of a thrombus forming 60 nm UFP dose. Although high doses were used, these results are consistent with the concept that (*i*) UFP, but not fine particles, translocate from the lung to the blood circulation and can affect endothelial function; (*ii*) that surface properties – including charge and also chemistry (20, 67) are important determinants of UFP effects.

UFP translocation to the blood circulation could also explain epidemiological findings of cardiovascular effects associated with ambient UFP and results of clinical studies showing vascular responses to inhaled ultrafine elemental carbon particles (1, 6, 13). As discussed before, evidence in humans for the translocation of inhaled UFP into the blood circulation is ambiguous, with one study showing rapid appearance in the blood and significant accumulation of label in the liver of humans inhaling ^{99}Tc-labelled 20 nm carbon particles (14), while another study using the same labeled particles reported no such accumulation (23). As described in the previous section, translocation of inhaled ultrafine particles in rats to extrapulmonary organs (liver) has been observed, albeit the magnitude of such translocation seems to be dependent on particle chemistry. Taken together all of the evidence from animal and human studies for alveolar translocation of UFP, it is likely that it happens in humans as well, dependent on particle surface characteristics/ chemistry which seems to have a significant influence on extrapulmonary translocation of UFP. In addition to UFP translocation, cardiovascular effects may also be the corollary of a sequence of events starting with particle-induced alveolar inflammation which in turn causes acute changes in blood coagulability and leading to cardiovascular effects (68).

Little recognized as a mechanism of importance in the area of UFP health effects and even more intriguing are observations of solid UFP translocation from airway mucosa along sensory nerves. Questions about the relevance for humans and about mechanisms of neuronal uptake and transport arise. It has been well demonstrated in rodents in numerous studies that inhaled or intranasally instilled soluble metal compounds are translocating *via* olfactory neurons to the olfactory bulb (69–72). No data in that regard are available for humans; however, the aforementioned results in chimpanzees and squirrel monkeys with intranasally-instilled 30 nm virus and especially 50 nm gold particles (57–59) clearly demonstrated in detail the uptake and movement of these particles in olfactory rods, olfactory dendrites and axons, and even crossing synapses in the olfactory bulb in non-human primates. Although this makes it a very likely mechanism in humans as well, there may also be important differences. Given that the olfactory mucosa of the human nose comprises only 5% of the total nasal mucosal surface as opposed to 50% in rats —which in addition are obligatory nose breathers—one can argue that the olfactory route may be an important transfer route to the CNS for inhaled UFP in animals with a well-developed olfaction system yet at the same time its importance for humans with a more rudimentary olfactory system can be questioned. The following comparison between rats and humans is an attempt to analyze the plausibility of human CNS exposure from inhaled UFP *via* the olfactory neuronal pathway.

The airflow directed to the human nasal olfactory mucosa (5% of total nasal mucosa) is 10% of the total nasal airflow (73). A much larger relative olfactory mucosa of 50% of the nasal mucosa in the rat receives only 15% of the total nasal airflow (74), so that the relative deposited amount per unit surface area of the olfactory mucosa may be not that different between rats and humans. Indeed, the Multiple Path Particle Deposition (MPPD; 36) model predicts that deposition of inhaled 20 nm particles in the nasopharyngeal region is about 5 times greater per unit nasal surface area in humans than in rats, i.e., ~60 ng/cm^2 for rats compared to ~300 ng/cm^2 in humans, assuming a 6-hr. exposure at 100 μg/m^3 under normal resting breathing conditions for both species. Assuming even distribution across the nasal mucosa—which is most likely not quite correct—rats would deposit ~480 ng on their olfactory mucosa and humans about ~1,575 ng. A first attempt to estimate how much of the deposit on the olfactory mucosa translocates to the olfactory bulb was made in our latest rat inhalation study with inhaled ultrafine carbon ^{13}C (Fig. 5). It was found that ~20% of the deposited amount translocated to the olfactory bulb within 7 days after the exposure. In the example above the rats would translocate 96 ng to their olfactory bulbs (85 mg organ weight), and humans would translocate 315 ng to their olfactory bulbs (168 mg organ weight [75]); that is 1.1 ng/mg olfactory bulb in rats and 1.9 ng/mg olfactory bulb in humans, a 1.6-fold higher concentration in humans. This is a rough estimate only and should not be over-interpreted and generalized for all UFP since neuronal translocation rates may be very dependent on particle size, surface chemistry and other parameters, and may also be different between rodents and humans.

If – instead of assuming even distribution of deposited ultrafine particles – one takes a different approach and assumes that inhaled 20 nm particles are depositing on the olfactory mucosa proportionately to the airflow directed to that region, the result changes even more towards a higher surface area dose in the human olfactory mucosa and a lower one in rats. In rats it would be 20 ng/cm^2 (160 ng on total olfactory mucosa), and in humans 600 ng/cm^2 (3150 ng on total olfactory mucosa), a 30-fold difference. The olfactory bulb would then be dosed with 32 ng in rats and 630 ng in humans (20% translocation of deposited amount), equivalent to olfactory bulb concentration of 0.4 ng/mg in rats and 3.8 ng/mg in humans, an almost 10-fold higher concentration in humans.

Of course, these modeling exercises do not prove that efficient olfactory translocation of inhaled solid UFP in humans does occur, but the evidence in non-human primates together with these dosimetric arguments strongly support the existence of this mechanism in humans despite a rudimentary olfaction system. Final confirmation obviously can only come from human evidence. However, even if the surface loading of the human olfactory mucosa is much less than the model predicts, and therefore the amount being translocated is much less, an exposure over many years or decades conceivably could result in significant accumulation in the olfactory bulb. Translocation into deeper brain structures is likely to occur as well, as shown in rats for soluble manganese (70), but requires further confirmatory studies with respect to solid ultrafine particles.

In addition to the olfactory system other neuronal translocation pathways in the respiratory tract need to be considered also, as discussed in the previous section. It is not only the olfactory mucosa that is supplied with sensory nerve endings, but the rest of the nasal and oro-pharyngeal mucosa have sensory nerves derived from the maxillary and ophthalmic branches of the trigeminus nerve, coming together at the trigeminal ganglion and continuing to the base of the brain. There is also a dense network of sensory nerves in the mucosa of the tracheobronchial region. These are additional neuronal translocation pathways for solid UFP and perhaps fine particles as Hunter and Dey (62) and Hunter and Undem (63) demonstrated in rodents. No evidence is available that these two additional particle translocation routes operate in humans (or non-human primates), it is a wide open field for future research. It can be hypothesized, however, that cardiovascular effects associated with ambient particles in epidemiological studies (76) are in part due to direct effects of translocated UFP in functions of the autonomic nervous system.

The chemistry of ambient UFP changes depend on particle size and source of origin. The smaller ambient ultrafine particles approaching the nanoparticle range consist increasingly of organic compounds (77), which are lipid or water soluble. It is conceivable that these, too, could be translocated via neurons since neuronal transport of proteins, lipids, and cellular organelles is a well-known phenomenon (78). A likely mechanism for neuronal transport of solid UFP is the cell's shuttle system of axonal and dendritic microtubuli (79), involving binding proteins of the Kinesin Superfamily. In the context of potential CNS effects of air pollution, including ambient UFP, a study by Calderon et al. (80) may point to an interesting link: These authors describe significant inflammatory changes of the olfactory mucosa, olfactory bulb and cortical and subcortical brain structures in dogs from a heavily polluted area in Mexico City, whereas these changes were not seen in dogs from a less polluted control city. However, whether direct effects of airborne UFP are the cause of these effects remains to be determined.

Outlook

The ubiquitous occurrence of airborne UFP results in significant human exposures under environmental and certain occupational conditions. Once deposited, the disposition of these particles appears to be unique: In addition to the classical clearance processes known to exist for fine and coarse particles, solid ultrafine particles can translocate to extrapulmonary organs across epithelial layers and also through uptake via sensory nerve endings in the nasal and tracheobronchial regions. Particularly, translocation of nasally deposited UFP along olfactory nerve axons into the olfactory brain has been well demonstrated in non-human primates and rodents, thereby circumventing the tight blood-brain barrier for accessing CNS structures. It is very likely that UFP reaching the CNS *via* this neuronal pathway will cause adverse effects, depending on their chemical composition and bioavailability of particle components to neuronal cells. Although this pathway is likely to exist in humans as well, conclusive evidence is still lacking. Thus, there is a need to determine the potential of airborne UFP to cause adverse CNS effects in laboratory animals and in humans. Figure 6 summarizes hypotheses of UFP effects, including systemic effects induced by alveolar inflammation or through particles reaching the blood circulation leading to cardiovascular effects; indirect cardiovascular effects *via* 'the autonomic nervous system from neuronally translocated particles to respective ganglia and CNS structures; and potentially direct effects on CNS functions. Modifying factors for these events most likely include age, underlying disease and other co-pollutants. Future research addressing these hypotheses requires close collaborations between toxicologists (animal, cellular, molecular), epidemiologists, clinicians (pulmonary, cardiovascular, neurologic), and atmospheric scientists.

Acknowledgement

This article is in part based on studies supported by an EPA Particulate Matter Center Grant (R-827354) and NIEHS Center Grant (P30ESO1247). This paper is based on parts of a review article prepared for publication in *Environmental Health Perspectives*

References

1. Wichmann, H-E., Spix, C., Tuch, T., Wölke, G., Peters, A., Heinrich, H., Kreyling, W.G., Heyder, J. (2000) *HEI Research Report No. 98.*

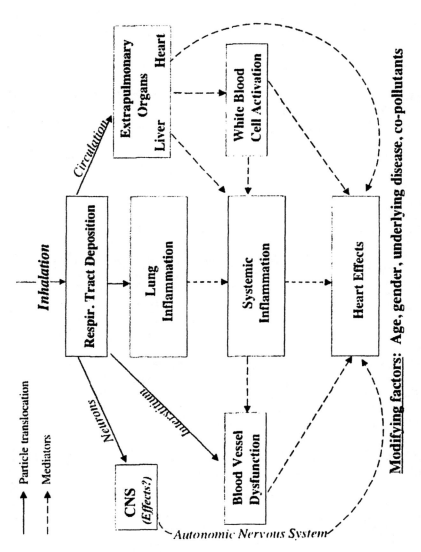

Figure 6: Potential mechanisms of effects of inhaled ultrafine particles.

2. Peters, A, Wichmann, HE, Tuch ,T, Heinrich, J, Heyder, J (1997a). *Am. J. Resp. Crit. Care Med.* 155: 1376-1383.

3. Peters, A, Doring, A, Wichmann, H-E, and Koenig, W (1997b). *The Lancet* 349 (No. 9065): 1582-1587.

4. Penttinen, P, Timonen, KL, Tiittanen, P, Mirme, A, Ruuskanen, J, Pekkanen, J (2001) *European Respiratory Journal* 17 (No. 3): 428-435.

5. Klot, von S., Wölke, G., Tuch, T., Heinrich, J., Dockery DW, Schwartz, J., Kreyling, WG, Wichmann, HE, Peters, A. (2002*Eur. Respir. J.* 20: 691-702.

6. Pekkanen, J, Peters, A, Hoek, G, Tiittanen, P, Brunekreef, B, *et al.* (2002) *Circulation* 106: 933-938.

7. Pekkanen,J., Timonen,K.L., Ruuskanen,J., Reponen,A., Mirme,A. (1997) *Environmental Research* 74: 24-33.

8. Tiittanen, P., Timonen, K.L., Ruuskanen, J., Mirme, A., Pekkanen, J. (1999). . *European Respiratory Journal* 13(2): 266-273.

9. Anderson, P.J., Wilson, J.D., Hiller, F.C. (1990) *Chest* 97: 1115-1120.

10. Daigle, C. C., Chalupa, D. C., Gibb, F. R., Morrow, P.E., Oberdörster, G., Utell, M.J., and Frampton, M.W. *Inhalation Toxicology* 15: 539-552, 2003.

11. Brown, J.S., Zeman, K.L., and Bennett, W.D. 2002.. *Am. J. Respir. Crit. Care Med.* 166: 1240-1247.

12. Kim, C.S., Jaques, P.A. (2000). *Phil. Trans. Roy. Soc. London A* 358: 2693-2705.

13. Pietropaoli AP, Frampton MW, Hyde RW, Morrow PE, Oberdörster G, Cox C, Speers DM, Frasier LM, Chalupa DC, Huang L-S, Utell MJ.(2004) *Inhalation Toxicol*, in press.

14. Nemmar, A, Hoet, PHM, Vanquickenborne, B, Dinsdale, D, Thomeer, M. et al. (2002a). *Circulation* 105: 411-414.

15. Nemmar, A., Hoylaerts, M.F., Dinsdale, D., Smith, T., Xu, H., Vermylen, J., Nemery, B. 2002b. *Am. J. Respir. and Critical Care Med.* 166: 998-1004.

16. Elder, ACP, Gelein R, Finkelstein JN, Cox C and Oberdörster G (2000). *Inhalation Toxicology* 12 (Suppl. 4): 227-246

17. Elder, AEP, R. Gelein, M. Azadniv, M. Frampton, J. Finkelstein, and G.Oberdörster. *Annals of Occup. Hygiene* 46 (Suppl 1): 231-234, 2002

18. Elder, ACP, Gelein, R, Azadniv, M, Frampton, M, Finkelstein, J, Oberdörster, G (2004). *Inhalation Toxicology,* in press.

19. Zhou, Y-M., Zhong, C-Y., Kennedy, I.M., *et al.* (2003). *Toxicology & Applied Pharmacology* 190: 157-169.

20. Oberdörster G, Finkelstein JN, Johnston C, Gelein R, Cox C, Baggs R, Elder ACP. Research Report No. 96, August, 2000.

21. Kreyling,WG, Semmler, M, Erbe, F, Mayer, P, Takenaka, S, Schulz, H, Oberdörster, G, and Ziesenis, A (2002). *J. Tox. & Environ. Health* 65 (20): 1513-1530.

22. Li, N., Sioutas, C., Cho, A., Schmitz, D., Misra, C., *et al.* (2003). Environmental Health Perspectives 111 (No. 4): 455-460.
23. Brown, D.M., Wilson, M.R., MacNee, W., Stone, V., Donaldson, K. (2001). *Toxicology & Applied Pharmacol.* 175: 191-199.
24. Donaldson, K., Brown, D., Clouter, A., Duffin, R., Macnee, W., *et al.* (2002). *J. Aerosol Medicine* 15 (2): 213-220.
25. Finkelstein, J.N., Reed, C.M., Johnston, C.J., Elder, A.C., Oberdörster, G. (2003). Alveolar macrophage production of inflammatory cytokines induced by ultrafine particles is increased in aged animals. Presented at the 2003 AAAR PM Meeting (Particulate Matter: Atmospheric Sciences, Exposure and the Fourth Colloquium on PM and Human Health), Pittsburgh, PA.
26. ICRP, *Annals of the ICRP: Human Respiratory Tract Model for Radiological Protection.* ICRP Publication 66, H. Smith, ed., Vol. 24, Nos. 1-3, Pergamon, 1994.
27. EPA (1996). Air quality criteria for particulate matter. Research Triangle Park, NC: National Center for Environmental Assessment-RTP Office, Report No. EPA /600/P95/001aF-cF.3v.
28. Swift, DL, Montassier, N, Hopke, PK, Karpen-Hayes, K, Cheng, Y-S, *et al.* (1992). *J. Aerosol Sci.* 23: 65-72.
29. Cheng, Y.S., Yeh, H.C., Guilmette, R.A., Simpson, S.Q., Cheng, K.H., Swift, D. L. (1996). Aerosol Science & Technology 25: 274-291.
31. Schiller, Ch.f., Gebhart, J., Heyder, J. Rudolf, G., Stahlhofen, W. (1988). *Ann. occup. Hyg.* 3 (Suppl. 1): 41-49.
32. Jaques, P.A., Kim, C.S. (2000). *Inhalation Toxicology* 12: 715-731.
34. Balaschazy, I., Hofmann, W., Heistracher, T. (1999) *J. Aerosol Sci.* 30 (No. 2): 185-203.
35. Balaschazy, I., Hofmann, W., Heistracher, T. (2003). *J. Applied Physiol.* 94: 1719-1725.
36. CIIT/RIVM (2002). Multiple Path Particle Model, MPPD, V1.0.
37. Kreyling, W. and Scheuch, G. (2000). Clearance of particles deposited in the lungs. In: Particle-Lung Interactions, Chapter 7, pgs. 323-376. P. Gehr, Heyder, J. (editors). Marcel Dekker, Inc., Publishers.
38. Schürch,S., Gehr,P., Hof,V.I., Geiser, M., Green,F. (1990). *Respiration Physiology* 80: 17-32.
39. Stahlhofen, W., Scheuch, G., Bailey, M.R. (1995). . *Radiation Protection Dosimetry* 60 (4): 311-319.
40. Warheit,D.B., Overby,L.H., George,G., Brody,A.R. (1988). *Exp Lung Res* 14: 51 – 66.
41. Warheit, D.B., Hill, L.H., George, G., Brody, A.R. (1986 *Am. Rev. Resp. Dis.* 134: 128-133.
42. Warheit, D.B., Hartsky, M.A. (1993). *Microscopy Research & Technique* 26: 412-422.

43. Hahn, F.F., Newton, G.J., Bryant, P.L. 1977. *In vitro* phagocytosis of respirable-sized monodisperse particles by alveolar macrophages <u>In</u>: <u>Pulmonary Macrophages and Epithelial Cells</u>. Sanders *et al.*, eds., ERDA Symposium Series, pgs. 424-435.

44. Tabata Y., and Ikada, Y. (1988). *Biomaterials* 9: 356-362.

45. Green,T.R.. Fisher,J., Stone,M., Wroblewski, B.M., Ingham,E. 1998. *Biomaterials* 19: 2297-2302.

46. Stone, V., Shaw, J., Brown, D.M., MacNee, W., Faux, S.P. and Donaldson, K. (1998). *Toxicol. In Vitro* 12: 649-659.

47. Oberdörster, G., J. Ferin, R. Gelein, S.C. Soderholm and J. Finkelstein. *Environmental Health Perspectives* 97: 193-197, 1992.

48. Ferin, J., G. Oberdörster and D.P. Penney. *Am. J. Resp. Cell and Molecular Biology* 6: 535-542, 1992.

49. Nikula, K.J., Avila, K.J., Griffith, W.C., Mauderly, J.L. (1997. Fundamental and Applied Toxicology 37: 37-53.

50. ILSI Risk Science Institute Workshop: *Inhalation Toxicology* 12: 1-17, 2000.

51. Oberdörster, G., Morrow, P.E., Spurny, K. (1988). *Ann. occ. Hyg.* 32 (Suppl., <u>Inhaled Particles, VI</u>): 149-156.

52. Rosenbruch, M. (1990). *Schweiz.Arch.Tierheikl.* 132: 469-470.

53. Rosenbruch, M. and Krombach, F. (1992). Structural and functional hepatic changes after experimental long-term inhalation of silica in non-human primates. 1992 Intl. Conference, American Lung Assoc, American Thoracic Society, May 17-20, 1992, Miami, FL. (Intl. Conf. Supplement to *American Review of Respiratory Disease*, 145; No. 4 (part 2 of 2), A91, April, 1992)

54. Oberdörster, G, Sharp, Z, Atudorei, V, Elder A, Gelein, R, Lunts, A, Kreyling, W, Cox, C. (2002). *J. Tox. & Environ. Health* 65 (20): 1531-1543.

55. Oberdörster, G. and Utell, M.J. (2002). *Environmental Health Perspectives* 110 (No. 8): A440-A441.

56. Patton, J.S. (1996. *Advanced Drug Delivery Reviews* 19: 3-36.

57. Bodian, D. and Howe, HA (1941) *Bulletin of the Johns Hopkins Hospital*, Vol. LXIX (No. 2), pgs. 79-85.

58. Howe, HA and Bodian, D (1941). *Bulletin of the Johns Hopkins Hospital*, Vol. LXIX (No. 2), pgs. 149-182.

59. De Lorenzo, AJD (1970). The olfactory neuron and the blood-brain barrier. <u>In</u>: <u>Taste and Smell in Vertebrates</u>. J&A Churchill, publishers, pgs. 151-176. Wolstenholme, GEW and Knight, J., eds. (CIBA Foundation Symposium Series).

60. Adams . RJ and Bray, D. 1983. *Nature* 303: 718-720.

62. Hunter, DD and Dey RD (1998). *Neuroscience* 83 (No. 2): 591-599.

63. Hunter, DD, and Undem, BJ (1999) *Am. J. Respir. Crit. Care Med.* 159: 1943-1948.

64. Katz, LC, Burkhalter, A and Dreyer, WJ (1984). *Nature* 310: 498-500.
65. Oberdörster, G., Sharp, Z., Atudorei, V., Elder, A., Gelein, R., Lunts, A., Kreyling, W., Cox, C. *Inhalation Toxicology*, 2004.
66. Nemmar, A., Hoylaerts, M.F., Hoet, P.H.M., Vermylen, J., Nemery, B. 2003. *Toxicology and Applied Pharmacology* 186: 38-45.
67. Kato, T., Yashiro, T., Murata, Y., Herbert, D.C., Oshikawa, K., *et al.* (2003). *Cell Tissue Res.* 311: 47-51.
68. Seaton, A., MacNee, W., Donaldson, K., Godden, D. (1995). Particulate air pollution and acute health effects. *Lancet* 345: 176-178.
69. Tjälve, H. and Henriksson, J. (1999) *Neurotoxicology* 20 (2-3): 181-196.
70. Gianutsos,G., Morrow,G.R., Morris,J.B. 1997 *Fundamental & Applied Toxicology* 37: 102-105.
71. Dorman, D.C., Brenneman, K.A., McElveen, A.M., Lynch, S.E., Roberts, K.C., Wong, B.A. 2002 *J. Tox. & Env. Health* (Part A: Current Issues) 65 (No. 20): 1493-1511.
72. Brenneman, K.A., Wong, B.A., Buccellato, M.A., Costa, E.R., Gross, E.A., Dorman, D.C. 2000. Direct olfactory transport of inhaled manganese (^{54}MnCl$_2$) to the rat brain: Toxicokinetic investigations in a unilateral nasal occlusion model. *Toxicol. Appl. Pharmacol.* 169: 238-248
73. Keyhani, K, Scherer, PW, Mozell, M.M (1997). *J. Theor. Biol.* 186: 279-301.
74. Kimbell, J.S., Godo, M.N., Gross, E.A., Joyner, D. R., Richardson, R.B., Morgan, K.T. 1997. *Toxicol & App Pharmacol.* 145: 388-398.
75. Turetsky, B.I., Moberg, P.J., Arnold, S.E., Doty, R.L., Gur, R.E. (2003). *Am. J. Psychiatry* 160 (4): 703-708.
76. Pope. C.A., III (2000). *Environmental Health Perspectives* 108 (Suppl. 4): 713-723.
77. Kittelson, DV. (1998). *J. Aerosol Science* 29 (5,6): 575-588.
78. Grafstein, B. and Forman, D.S. (1980). *Physiological Reviews* 60 (No. 4): 1167-1283.
79. Hirokawa, N. 1998. *Science* 279: 519-526.
80. Calderon-Garciduenas, L., Azzarelli, B., Acune, H., Garcia, R., Gambling, T.M., *et al.* (2002). *Toxicologic Pathology* 30 (No. 3): 373-389.

Chapter 8

Toxicity of Single-Wall Carbon Nanotubes in the Lungs of Mice Exposed by Intratracheal Instillation

Chiu-wing Lam [1,2], John T. James[1], Richard McCluskey[1], and Robert L. Hunter[3]

[1]Space and Life Sciences, NASA Johnson Space Center, Houston, TX 77058
[2]Wyle Laboratories, 1290 Hercules Dr, Suite 120, Houston, TX 77058
[3]Department of Pathology and Laboratory Medicine, University of Texas at Houston, Medical School, Houston, TX 77030

Single-wall carbon nanotubes (NTs) are light and could become airborne; fine NT particles could potentially reach the lung. The pulmonary toxicity of three NT products was investigated in mice by intratracheal instillation. Mice (4 to 5 per group) were each instilled once with a fine-particle suspension containing 0, 0.1 or 0.5 mg of NTs or a reference dust (carbon black or quartz), and killed 7 or 90 days later for lung histopathological study. Carbon black elicited minimal effects and high-dose quartz produced mild to moderate inflammation in the lungs. All the three NT products studied produced granulomas (microscopic nodules) and other lung lesions. These results show that, for the test conditions described here and on an equal-weight basis, if NTs reach the lung, they can be more toxic than quartz. If airborne NT dusts are present in the working environment, respiratory protection should be used to minimize inhalation exposures.

Introduction

NTs structurally resemble rolled-up graphite sheets with one end capped. They possess unique electrical, mechanical, and thermal properties and are finding widespread applications in the electronics, computer, and aerospace industries. Dr. Richard Smalley, a Nobel laureate and a pioneer in NT research at Rice University (Houston, TX), has predicted that hundreds or thousands of tons of NTs could be produced in 5 to 10 years and "in time, millions of tonnes of nanotubes will be produced worldwide every year " *(1, 2)*. As the production and applications of NTs expand, the potential for human exposures will also increase.

NTs are commonly produced by electric arc or laser evaporation of carbon atoms from graphite, or by chemical-vapor deposition (CVD) using a non-graphite source. In each method, carbon atoms generated are allowed to deposit on vaporized catalytic metal particles forming carbon nanotubes. High-pressure CO conversion (HiPco™) is a CVD process that uses carbon monoxide as carbon source *(3)*. An individual NT molecule or fiber is about 1 nm in diameter and several micrometers long. NT fibers pack tightly into bundles or ropes of microscopic sizes. These bundles, in turn, aggregate loosely into small clumps. They are very light and can easily become airborne. A study by the National Institute of Occupational Safety and Health on a raw HiPco-NT sample showed that agitation could generate low concentrations of respirable-size particles in the air *(4)*.

Concern about the potential for its workers to be exposed to this novel and important material of unknown toxicity prompted NASA to sponsor the present NT pulmonary toxicity study. Reported here is a summary [(for the full report, see the December 2003 issue of *Toxicol. Sci.* *(5)*] of our investigation of the pulmonary toxicity of three NT products, generously provided by Rice University and CarboLex, Inc. (Lexington, KY). These products were made by different methods, and contained different types or amounts of metals. Metal analysis in our laboratory showed that the Rice HiPco™-prepared NTs contained about 25% (w/w) iron in the raw form (RNT), and 2% iron after purification (PNT); the CarboLex carbon-arc product (CNT) contained 25% nickel and 5% yttrium *(5)*. The CNT product is a fine powder like carbon toner. The RNT particles tend to stick to each other or to container walls; it would be very difficult to isolate enough particles of respirable sizes for an inhalation study, and to generate and monitor NT aerosols for such a study. Thus the NTs were tested for pulmonary toxicity by intratracheal instillation, which is a common and accepted route of administration for toxicologically screening dusts without incurring the expense and difficulty of an inhalation study (6, 7). Intratracheal instillation studies also allow comparative toxicity investigation of several dusts simultaneously *(8)*. In the present study, carbon black and quartz, two dusts whose toxicities in the lungs have been well characterized, were used as references.

Methods

To produce results that are relevant for assessing the toxicity of a dust in the lung, a test material should have particle sizes in the respirable range. The RNT and PNT, consisting predominately of large particles, are not suitable for lung toxicity study without further treatment. NTs are neither water soluble nor wettable. Walters *et al.,* (9) of Rice University, used "aggressive sonication of purified NT samples in surfactants such as Triton-X or highly polar solvents, like dimethyl formamide" to make suspensions (10 mg/L) containing mostly individual fibers and a few small bundles. We prepared fine particle suspensions suitable for intratracheal instillation by briefly shearing (2 min) and sonicating (30 sec) NT samples in mouse serum. R. Hauge of Rice (personal communication) advised us that brief sonication would not shorten or change the fundamental nature of NTs. NT sample suspensions prepared in our laboratory appeared as smooth pastes; under the microscope, they showed that most of the particles were less than 10 μm in size, but some larger aggregates were present. All test samples were prepared in the same way. Male mice (B6C3F$_1$, ca. 30 g, 4 to 5 per group) were each instilled once with 0, 0.1 or 0.5 mg of NTs, carbon black, or quartz suspended in 50 μl of heat-inactivated mouse serum. Seven or 90 days after a single dust treatment, the mice were euthanized and the lungs were excised, fixed, and stained for histopathological study.

Results and Discussion

As expected, the serum control group showed no signs of pulmonary toxicity, and the lungs of the CB group showed black particles in alveolar regions with minimal changes (Figure 1A). The quartz-treated mice in the high-dose group showed mild to moderate pulmonary inflammation. All the NT products induced a dose-dependent response characterized by the formation of epithelioid granulomas (microscopic nodules) in the centrilobular alveolar septa and, in some cases, interstitial inflammation and necrosis in the lungs (Figures 1B and 1C). These lesions persisted and, in most cases, were more pronounced in the 90-d groups.

The fact that the purified NTs (PNTs), whose metals were removed by rigorous chemical treatment [prolonged reflux in concentrated acid (10)], produced prominent granulomas strongly indicates that NTs themselves induced a granulomatous reaction in the lung. Granulomas were also observed by Warheit of DuPont Company in a similar study, in which rats were intratracheally instilled with a laser-produced NT product (see Warheit's report in this book); the instilled samples, like ours, were sonicated in a dispersing agent. The NT dosage of 5 mg/kg that produced granulomas in rats in Warheit's study was comparable to that of our low-dose (0.1 mg/30-g mouse or 3.3 mg/kg) group in mice. At this dose (0.1 mg/mouse), we saw granulomas in the mice treated with the HiPco products (RNT and PNT), but not with the CNT. The lungs of all mice exposed to NTs at the higher dose (0.5 mg/lung) had granulomas and some other lesions such as interstitial inflammation.

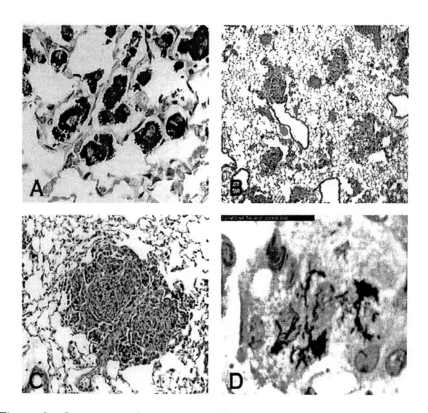

Figure 1. Lung tissue from mice instilled with 0.5 mg of carbon black or carbon nanotube material per mouse and euthanized 7 d or 90 d after the single treatment: (A) 7 d carbon black: black dust particles in macrophages scattered in alveoli, (B) 7 d RNT: NT-induced granulomas in the lungs shown in low magnification, (C) 7 d PNT: a close-up view of a granuloma containing black particles laden in macrophages, (D) 90 d PNT: a granuloma containing NT fibers shown at a high magnification. Magnifications varied from 40 to 900×.

Although both types of carbon particles (CB and NTs) were taken up by alveolar macrophages, their fate and reactions in the tissue were very different. CB-laden macrophages scattered in the alveolar space (Figure 1A), but NT-laden macrophages moved rapidly to centrilobular locations, where they entered alveolar septae and clustered to form epithelioid granulomas (Figures 1B and 1C). It is well known that if the lung is not overloaded with dust, dust-laden macrophages on the alveolar surface will migrate upward and be carried by the escalator/mucociliary system up the trachea, and eventually be cleared into the esophagus. However, when dusts enter the interstitial or subepithelial space, they are very difficult to clear from the lung. These findings indicate that NTs and CB have very different intrinsic toxicities and are subjected to different clearance mechanisms in the lung.

But what makes NTs and CB toxicologically different? In addition to their difference in surface chemistry (NTs are an excellent electrical conductor whereas CB is not), they are also different in geometry. Microscopically, the thin NT fibers pack tightly and in parallel to form ropes or rods (1). As defined in *Comprehensive Toxicology,* "Fibers are a special class of particles defined as elongated objects whose aspect ratio, the ratio of the object's length to its diameter, is greater than three" (11). Individual NT molecules and assembled NT ropes, rods, or bundles are fibers, whereas CB is amorphous. The fibrous appearance of NTs is clearly visible in the granulomas (Figure1D). Bio-persistent fibers are a class of particulate that is of toxicological concern (12).

Generally, insoluble dust particles on alveolar surfaces are phagocytized by macrophages, and the dust-laden macrophages are cleared by the mucociliary clearance mechanism. However, if the macrophages enter the interstitial space, dust particles in the interstitial septa cannot be removed from the lung by the mucociliary clearance mechanism. This is the case with NTs. Because we see a difference in how alveolar macrophages react to CB and NT particles, it is reasonable to assume that the particle clearance depends on the properties of the dust, and not the route of exposure. In other words, macrophages will likely react similarly to NT particles introduced into the lung by an intratracheal instillation or by inhalation exposures. Under either of these conditions, the lung burden of NTs from an intratracheal dose could be attained by repeated inhalation exposures (concentration × time), and therefore be used for rough estimation of inhalation exposure concentrations and duration. It has been estimated that the fraction of inhaled dust that is deposited deep in the human lung is about 30% for 3-μm particles and increases to 55% for 0.05-μm particles (13). If we assume that 40% of the inhaled respirable NT particles deposit in the pulmonary region, and that a 30-g mouse breathes in 30 ml of air per min (14), then a mouse breathing respirable NT dust at 5 mg/m^3 for 8 h daily would accumulate 0.029 mg NTs/day. After breathing the same concentration of NTs for 3.5 or 17 days, a mouse would accumulate 0.1 mg or 0.5 mg of NTs in the lung, respectively. The exposure concentration of 5 mg/m^3 is the same concentration as the Permissible Exposure Limit (PEL) that OSHA set for respirable synthetic graphite dust and is also the exposure limit recommended for SWNTs by a manufacturer and supplier (*15*).

OSHA's PELs, which are time-weighted (40 h/wk) concentrations, are set for lifetime occupational exposures. It can be inferred from the exposure-risk estimation information presented here that if workers were chronically exposed to respirable NT dust, even at a fraction of the PEL for synthetic graphite, they would be likely to have serious lung lesions. Therefore the PEL for synthetic graphite must not be used for NTs. It also can be concluded that, for the test conditions described here and on an equal-weight basis, if NTs reach the lungs, they are much more toxic than carbon black and can be more toxic than quartz, which is considered a serious occupational health hazard in chronic inhalation exposures. If fine NT dusts are present in a work environment, exposure protection strategies should be implemented to minimize human exposures.

References

1. ISI Nanotechnology. An interview with Dr. Richard Smalley. ISI Essential Science Indicators Special Topics, 2002. http://www.esi-topics.com/nano/interviews/Richard-Smalley.html
2. Ball, P. Focus on carbon nanotubes. *Nature Scienceupdate*, 1999. http://www.nature.com/nsu/991202/991202-1.html
3. Bronikowski, M. J.; Willis, P. A.; Colbert, D. T.; Smith, K. A.; and Smalley, R. E. Gas-phase production of carbon single-walled nanotubes from carbon monoxide via the HiPco process: a parametric study. *J. Vac. Sci. Technol.* **2001,** *A* 19, 1800-5.
4. Baron, P. A.; Maynard, A. D.; and Foley, M. Evaluation of aerosol release during the handling of unrefined single walled carbon nanotube material. National Institute of Occupational Safety and Health, Cincinnati, OH, NTIS PB2003-102401, 2003.
5. Lam, C.-W.; James, J. T.; McCluskey R.; and Hunter, R.L. Pulmonary toxicity of carbon nanotubes in mice 7 and 90 days after intratracheal instillation. *Toxicol. Sci.* **2003,** Vol. 76 (in press).
6. Leong, B. K.; Coombs, J. K.; Sabaitis, C. P.; Rop, D. A.; and Aaron, C. S. Quantitative morphometric analysis of pulmonary deposition of aerosol particles inhaled via intratracheal nebulization, intratracheal instillation or nose-only inhalation in rats. *J. Appl. Toxicol.* **1998,** 18, 149-60.

66

7. Driscoll, K. E.; Costa, D. L.; Hatch, G.; Henderson, R.; Oberdorster, G.; Salem, H.; and Schlesinger, R. B. Intratracheal instillation as an exposure technique for the evaluation of respiratory tract toxicity: uses and limitations. *Toxicol. Sci.* **2000**, 55, 24-35.
8. Lam, C.-W.; James, J. T.; McCluskey, R.; Cowper, S.; and Muro-Cacho, C. Pulmonary toxicity of simulated lunar and Martian dusts in mice. I. Histopathology 7 and 90 days after intratracheal instillation. *Inhal. Toxicol.* **2002**, 14, 901-16.
9. Walters, D. A.; Casavant, M. J.; Qin, X. C.; Huffman, C. B.; Boul, P. J.; Ericson, L. M.; Haroz, E. H.; O'Connell, M. J.; Smith, K.; Colbert, D. T.; and Smalley, R. E. In-plane-aligned membranes of carbon nanotubes. *Chem. Phys. Lett.* **2001**, 338, 14-20.
10. Rinzler, A.G.; Liu, J.; Dai. H.; Nikolaev, P.; Huffman, C.B.; Rodriguez-Macias, F.J.; Boul., P. J.; Lu, A. H.; Heymann, D.; Colbert, D.T.; Lee., R. S.; Fischer., J. E.; Rao, A. M.; Eklund, P.C.; and Smalley, R. E. Large-scale purification of single-wall carbon nanotubes: process, product, and characterization. *Appl. Phys.* **1998**, *A 67*, 29-37.
11. McClellan, R. Nuisance dusts. In: *Toxicology of the Respiratory System.* Sipes, I. G.; McQueen, C. A.; and Gandolfi, A. J. Eds., Vol. 8. Elsevier Science Ltd, New York, NY. 1997.
12. Oberdorster, G. Determinants of the pathogenicity of man-made vitreous fibers (MMVF). *Int. Arch. Occup. Environ. Health,* **2000**, 73 Suppl, S60-8.
13. Bates, D. V.; Fish, B. R.; Hatch, T. F.; Mercer, T. T.; Morrow, P. E. Deposition and retention models for internal dosimetry of the human respiratory tract. *Hlth. Phys.* **1966**, *12*, 173.
14. R. A. Parent. *Comparative Biology of the Normal Lung.* CRC Press, Boca Raton, FL. 1992.
15. Material Safety Data Sheet (MSDS) on Nanotubes, Carbon Nanotechnology Inc., Houston, TX, USA.

Acknowledgements

The authors thank the following colleagues for their advice or technical assistance: A. Lee, T. Blasdel, B. Leong, G. Oberdörster, A Holian, J. Nelle, S. Bassett, M. Kuo, J. Read, H. Garcia, J. Krauhs, B. Conaway, and staff of the Center for Nanoscale Science and Technology (CNST) and Center for Biological and Environmental Nanotechnology at Rice University. The generous gifts of carbon nanotube samples from Rice CNST and from CarboLex Inc., Printex 90 from Degussa Corp. and Mil-U-Sil-5 from US Silica are acknowledged. This study was supported by a NASA grant.

Chapter 9

Lung Toxicity Bioassay Study in Rats with Single-Wall Carbon Nanotubes

D. B. Warheit[1], B. R. Laurence[1], K. L. Reed[1], D. H. Roach[2], G. A. M. Reynolds[2], and T. R. Webb[1]

[1]DuPont Haskell Laboratory for Health and Environmental Sciences, Newark, DE 19714
[2]Central Research and Development, DuPont, Wilmington, DE 19880

The objective of this study was to evaluate the toxicity of intratracheally instilled single wall carbon nanotubes (SWCNT) in the lungs of rats. The pulmonary toxicity of intratracheally instilled carbon nanotubes was compared with a positive control particle-type, quartz, and carbonyl iron particles (a negative control particle-type). The lungs of rats were intratracheally instilled with either 1 or 5 mg/kg of the following control or particle-types: 1) carbon nanotubes; 2) quartz particles, 3) carbonyl iron particles. Phosphate-buffered saline (PBS) and PBS + 1% Tween 80 instilled rats served as additional controls. After exposures, the lungs of PBS and particle-exposed rats were assessed both using bronchoalveolar lavage (BAL) fluid biomarkers, cell proliferation methods, and by histopathological evaluation of lung tissue at 24 hrs, 1 week, 1 month and 3 months post-instillation exposure.

High dose exposures to SWCNT produced mortality in ~15% of instilled rats within 24 hrs post-instillation. This mortality was due to mechanical blockage of the large airways by the instillate, and did not result from inherent lung toxicity of the instilled SWCNT particulate.

Data from the bronchoalveolar lavage and cell proliferation studies demonstrated that lung exposures to quartz particles produced persistent enhancement in pulmonary inflammation, cytotoxicity, and lung cell parenchymal cell proliferation indices. Alternatively, single wall carbon nanotube exposures produced transient inflammatory and cell injury effects at 1 day postexposure, due primarily to the blockage of airways and resulting injury by the instillate.

Histopathological analyses revealed that exposures to quartz particles (5 mg/kg) resulted in dose-dependent lung inflammatory responses, in association with accumulation of foamy alveolar macrophages and early lung fibrosis at the sites of normal particle deposition. Pulmonary exposures to carbonyl iron particles produced no significant adverse effects. Lung exposures to SWCNT in rats produced a non dose-dependent foreign tissue body reaction, as evidenced by a series of multifocal mononuclear granulomas. The granulomas were characterized by black SWCNT bolus material in the center, and surrounded by macrophage-like giant cells. The lesions did not appear to progress beyond 1 month postexposure. Surprisingly, the bronchoalveolar lavage and cell proliferation results were not predictive biomarkers of the SWCNT-induced granulomatous lesions, unlike pulmonary responses to quartz particles.

The observation of a SWCNT dust-induced foreign tissue reaction is not consistent with the following: 1) lack of lung toxicity by assessing lavage parameters; 2) lack of lung toxicity by measuring cell proliferation parameters; 3) an apparent lack of a dose response relationship; 4) non-uniform distribution of lesions; 5) the paradigm of dust-related lung toxicity effects; 6) possible regression of effects over time. Moreover, recently reported data from two exposure assessment studies at the workplace indicate very low aerosol SWCNT exposure levels. Therefore, the physiological relevance of these findings remains to be determined. Thus, to reconcile the apparent discrepancies in this lung bioassay study, it is critical that the pulmonary effects of SWCNT soot in rats be assessed by generating SWCNT aerosols in an inhalation toxicity study.

Introduction

Carbon nanotubes have excellent mechanical, electrical and magnetic properties. Single-wall nanotubes (SWCNT), due to electrostatic properties, self-organize into rope-like structures that can range in lengths up to several microns. The potential hazards of inhalation exposure to carbon nanotubes have not been elucidated. Recent experimental studies in rats indicate that inhaled, nano-sized carbon black particles may produce significant lung toxicity in rats and the toxicity potential is inversely related to particle size. Thus, nano-sized carbon black particles produce enhanced lung toxicity in rats when compared to larger-sized carbon black particles (furnace black particles – mean diameters = 14 nm (surface area = 270 m^2/g); lamp black particles – mean diameters = 95 nm (surface area = 22 m^2/g) (1,2,3).

This study was designed as a hazard screen to assess whether SWCNT particulate exposures produce significant toxicity in the lungs of rats by comparing the activity of the carbon-derived particulates with other reference particulate materials. Thus, the objective was to evaluate in rats, using a well-developed, short-term lung bioassay, 1) the pulmonary toxicity impact of intratracheally instilled SWCNT samples and to compare the pulmonary effects with a low and high toxicity particulate samples.

Methods

General Experimental Design (see Tables 1-2; Fig. 1)

The major features of the pulmonary bioassay bridging study are 1) dose response evaluation, and 2) time course assessments to determine the persistence of any observed effect. Thus, the major endpoints of this study were the following: 1) time course and dose/response intensity of lung inflammation and cytotoxicity; 2) alveolar macrophage chemotactic function at 1 week post-instillation exposure; 3) airway and lung parenchymal cell proliferation indices; and 4) histopathological evaluation of lung tissue.

Groups of rats were intratracheally instilled with 1 or 5 mg/kg of carbon nanotubes (CNT), quartz-crystalline silica particles (Q) , or carbonyl iron (CI) particles. All particles were prepared in a volume of 1.0 % Tween 80 and phosphate-buffered saline (PBS) and subjected to polytron dispersement. Groups of PBS and PBS-Tween instilled rats served as controls. The lungs of PBS, PBS-Tween and particle-exposed rats were evaluated by bronchoalveolar lavage at 24 hr, 1 week, 1 month and 3 months postexposure (pe). For the morphological studies, additional groups of animals were instilled with the particle-types listed above as well as PBS and PBS-Tween, and graphite particles. These studies were dedicated for lung tissue analyses and consisted of

Table I. Experimental Groups for SWCNT Pulmonary Bioassay Study

Group	Description	Purpose	Dose (mg/kg)
1	Phosphate Buffered Saline (PBS)	Negative Control	
2	PBS + 1% Tween-80	Surfactant Control	
3	Carbon nanotubes (SWNT) + 1% Tween-80	Test Compound	1 and 5
4	Crystalline silica particles (Quartz) + 1% Tween-80	Positive Particle Control	1 and 5
5	Carbonyl iron particles + 1% Tween-80	Negative Particle Control	1 and 5

Table 2. Protocol for Carbon Nanotube Bioassay Study

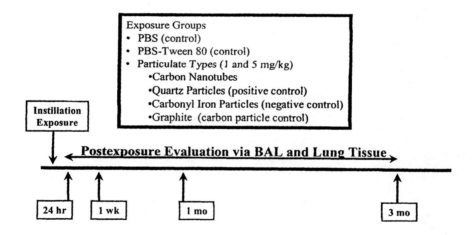

<u>Pulmonary Bioassay Bridging Studies</u>

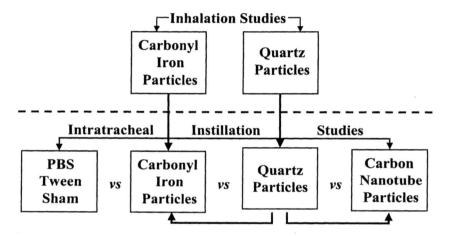

Figure 1.   Schematic demonstrating the strategy for conducting pulmonary bioassay bridging studies. Bridging studies can have utility in providing an inexpensive preliminary safety screen when evaluating the hazards of new developmental compounds. The basic idea for the bridging concept is that the effects of the instilled material serve as a control (known) material and then are "bridged" on the one hand to the inhalation toxicity data for that material, and on the other hand to the new materials being tested.

cell proliferation assessments and histopathological evaluations of the lower respiratory tract.

For the bronchoalveolar lavage studies, groups of male rats were exposed via intratracheal instillation to 1) vehicle control - Phosphate-buffered saline (PBS); 2) additional vehicle control - PBS + 1%Tween 80; 3) single wall carbon nanotubes in 1% Tween + PBS at 1 and 5 mg/kg; 4) Min-U-Sil crystalline quartz particles in 1% Tween + PBS at 1 and 5 mg/kg; 5) carbonyl iron particles in 1% Tween + PBS at 1 and 5 mg/kg. All particle-types were suspended in phosphate-buffered saline and 1% Tween 80.

For the morphological studies, additional groups of animals were instilled with the particle-types listed above plus graphite particles and the vehicle controls, i.e., PBS and PBS-Tween. These studies were dedicated to lung tissue analyses and consisted of cell proliferation assessments and histopathological evaluations of the lower respiratory tract. Similar to the BAL fluid studies, the intratracheal instillation exposure period was followed by 24-hour, 1-week, 1-month, and 3-month recovery periods.

Animals

Groups of male Crl:CD®(SD)IGS BR rats (Charles River Laboratories, Inc., Raleigh, North Carolina), approximately 8 weeks old at study start (mean weights in the range of 240 – 255 grams) were used in this study. All procedures using animals were reviewed and approved by the Institutional Animal Care and Use Committee (Haskell Animal Welfare Committee) and the animal program is fully accredited by the Association for Assessment and Accreditation of Laboratory Animal Care (AAALAC).

Particle-types

Single wall carbon nanotube soot generated via a laser ablation process (4), was obtained from GAM Reynolds and DH Roach of the DuPont Central Research. The nominal size of single wall nanotubes are 1.4 nm diameter x >1 um length. However, the nanotubes rarely exist as individual units and exist primarily as agglomerated "ropes" of nanotubes of ~30 nm in diameter. The soot is comprised of about 30-40 weight % amorphous carbon and 5 weight % each of nickel and cobalt with the balance being the carbon nanotube agglomerates. Quartz particles in the form of crystalline silica (Min-U-Sil 5) were obtained from Pittsburgh Glass and Sand Corporation. The particle sizes range from 1 – 3 μm. Carbonyl iron particles were obtained from GAF Corp. The particle sizes range from 0.8 μm to 3.0 μm.

Pulmonary Lavage

The lungs of sham and particulate-exposed rats were washed/lavaged with a warmed phosphate-buffered saline (PBS) solution. Techniques for cell counts, differentials and pulmonary biomarkers in lavaged fluids were utilized as previously described (*5,6*). All biochemical assays were performed on BAL fluids using a Roche Diagnostics (BMC)/Hitachi® 717 clinical chemistry analyzer using Roche Diagnostics (BMC)/Hitachi® reagents. Lactate dehydrogenase (LDH), alkaline phosphatase (ALP), and lavage fluid protein were measured using Roche Diagnostics (BMC)/Hitachi® reagents. Lactate dehydrogenase is a cytoplasmic enzyme and is a biomarker of cell injury. Alkaline phosphatase activity is an indicator of Type II lung epithelial cell cytotoxicity. Increases in BAL fluid protein concentrations generally are consistent with enhanced permeability of proteins migrating from the vasculature into alveolar regions.

Chemotaxis Studies

Macrophage chemotaxis activity represents an indicator of alveolar macrophage function. Chemotactic studies were conducted as previously described (*7*). Alveolar macrophages were collected from particulate-exposed or control rats by lavage as described above. The chemotaxis assay was carried out using three concentrations (i.e., 1, 5, and 10%) of normal heated sera as the chemotactic stimulus.

Pulmonary Cell Proliferation Studies

Following particle exposures, airway and lung parenchymal cell turnover in rats was measured following 24 hrs, 1 week, 1 or 3 month postexposure periods. Groups of particulate-exposed rats and corresponding controls were pulsed 24 hrs after instillation, as well as 1 week, 1 and 3 months postexposure, with an intraperitoneal injection of 5-bromo-2'deoxyuridine (BrdU) dissolved in a 0.5 N sodium bicarbonate buffer solution at a dose of 100 mg/kg body weight. The animals were euthanized 6 hrs later by pentobarbital injection. The lungs were then infused with a neutral buffered formalin fixative at a pressure of 21 cm H_2O. After 20 minutes of fixation, the trachea was clamped, and the heart and lungs were carefully removed *en bloc* and immersion-fixed in formalin. In addition, a 1-cm piece of duodenum (which served as a positive control) was removed and stored in formaldehyde. Subsequently, parasagittal sections from the right cranial and caudal lobes and regions of the left lung lobes as well as the duodenal sections were dehydrated in 70% ethanol and sectioned for histology. The sections were embedded in paraffin, cut, and mounted on glass slides. The slides were stained with an anti-BrdU antibody, with an AEC (3-amino-9-ethyl carbazole) marker, and counter-stained with aqueous hematoxylin. A minimum of 1000 cells/animal were counted each in terminal bronchiolar and alveolar regions by light microscopy at x 1000 magnification (5,6). The percentages of immunostained cells were quantified.

Morphological Studies

The lungs of rats exposed to particulate-exposed or PBS-Tween controls were prepared for microscopy by airway infusion under pressure (21 cm H_2O) at 24 hours, 1 week, 1 and 3 months postexposure. Sagittal sections of the left and right lungs were subsequently prepared for light microscopy (paraffin embedded, sectioned, and hematoxylin-eosin stained) and evaluated (5,6).

Statistical Analyses

For statistical analyses, each of the experimental values were compared to their corresponding sham control values for each time point. A one-way analysis of variance (ANOVA) and Bartlett's test were calculated for each sampling time. When the F test from ANOVA was significant, the Dunnett test was used to compare means from the control group and each of the groups exposed to particulates. Significance was judged at the 0.05 probability level.

Results

SWCNT-related Mortality

Intratracheal instillation exposures to high dose (5 mg/kg) single wall carbon nanotubes produced mortality in ~15% of the SWCNT-instilled rats within 24 hrs. After histopathological review, it was concluded that these deaths resulted from blockage of the major airways by the SWCNT instillate, and not from pulmonary toxicity per se of the instilled SWCNT particulate (Fig. 2). Thus, it is likely that the mortality caused by exposure to SWCNT 5 mg/kg was due to the exposure regimen, given the aggregating properties of the SWCNT which are highly electrostatic and do not separate into individual nanotubes (1

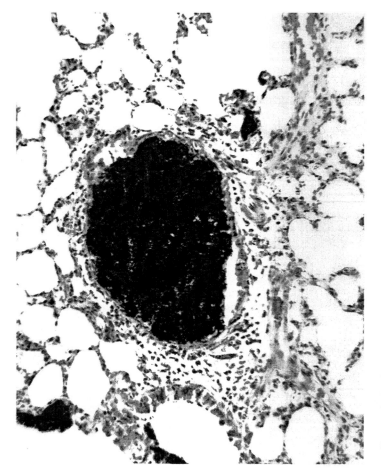

Figure 2. Light micrograph of lung tissue from a rat exposed to 5 mg/kg CNT (a few hours after exposure). The major airways are mechanically blocked by the CNT instillate. This led to suffocation in 15% of the CNT-exposed rats and was not evidence of pulmonary toxicity of CNT.

nm diameter x 1 μm length), but instead form "nanoropes" which consist of agglomerates of 10 – 100 nanotubes.

Lung Weights

Lung weights in 5 mg/kg SWCNT-exposed rats were increased vs. controls at 24 hrs, 1 week, and 1 month but not at 3 months postexposure (Fig. 3).

Bronchoalveolar Lavage Fluid Biomarker Results

Pulmonary Inflammation

The numbers of cells recovered by bronchoalveolar lavage from the lungs of high dose quartz-exposed (5 mg/kg) groups were significantly higher than any of the other groups for all postexposure time periods. Intratracheal instillation exposures to quartz particles (1 and 5 mg/kg) produced persistent lung inflammation, as measured through 3 months postexposure; in contrast, instillation of CI and SWCNT resulted in a transient, lung inflammatory response, as demonstrated by enhanced percentages of BAL-recovered neutrophils, measured at 24 hrs post exposure (Fig. 4).

BAL Fluid Parameters

Transient increases in BAL fluid lactate dehydrogenase values were measured in lung fluids of high dose (5 mg/kg) SWCNT-exposed rats at 24 hrs post exposure, but were not persistent. In contrast, exposures to 5 mg/kg quartz particles produced a sustained increase in BAL fluid LDH values relative to controls through the 3-month postexposure period (Fig. 5). Exposures to 5 mg/kg quartz particles produced a persistent increase in bronchoalveolar lavage fluid microprotein values at 24 hrs, 1 and 3 months postexposure, while transient increases in BAL fluid microprotein values were measured in the lung fluids recovered from high dose (5 mg/kg) SWCNT-exposed rats at 24 hrs postexposure, but were not different from control values at 1 week postexposure (Fig. 6). Transient increases in BAL fluid alkaline phosphatase values were measured only in the lung fluids of quartz-exposed rats at 24 hrs postexposure (1 mg/kg) and at the 1 week (5 mg/kg) postexposure time periods (Fig. 7).

The results of chemotaxis studies peformed at 1 week postexposure demonstrated that alveolar macrophages exposed to quartz particles (5 mg/kg) were impaired in their chemotactic responses to normal heated sera (NHS) when

78

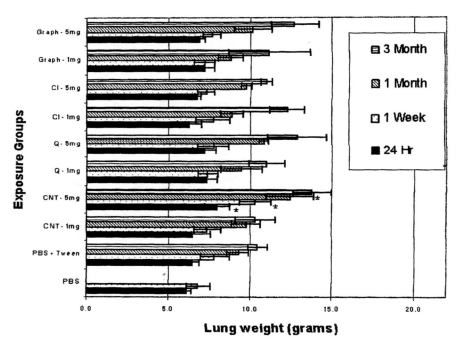

Figure 3. Lung weights of rats exposed to particulates and corresponding controls. Values given are group means ± S.D. Significant increases vs. controls were measured in the lungs of rats exposed to CNT 5 mg/kg at 24 hrs, 1 week and 1 month postexposure. *p < 0.05

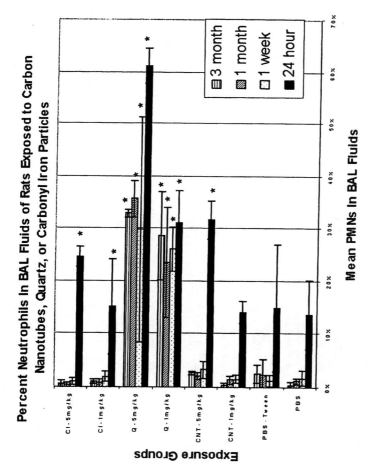

Figure 4. Pulmonary inflammation in particulate-exposed rats and controls as evidenced by % neutrophils (PMN) in BAL fluids at 24 hrs, 1 week, 1 month and 3 months postexposure (pe). Instillation exposures resulted in transient inflammatory responses for nearly all groups at 24 hrs pe. However, exposures to Quartz particles at 1 and 5 mg/kg produced a sustained lung inflammatory response. $p < 0.05$

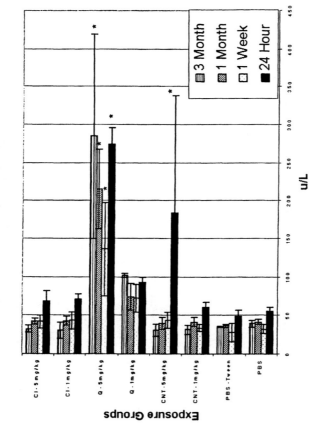

*Figure 5. BAL fluid LDH values for particulate-exposed rats and corresponding controls at 24 hrs, 1 week, 1 month and 3 months postexposure (pe). Significant increases in BALF LDH vs. controls were measured in the CNT 5 mg/kg exposed group at 24 hrs pe and the 5 mg/kg Quartz-exposed animals at all 4 time periods pe. *p < 0.05*

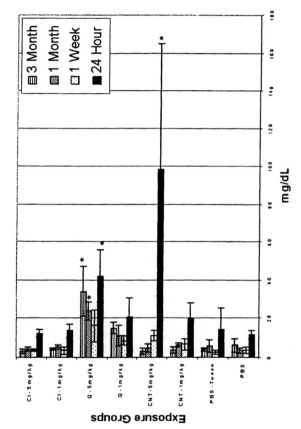

*Figure 6. BAL fluid protein (MTP) values for particulate-exposed rats and corresponding controls at 24 hrs, 1 week, 1 month and 3 months postexposure (pe). Significant increases in BALF MTP vs. controls were measured in the CNT 5 mg/kg exposed group at 24 hrs pe and the 5 mg/kg Quartz-exposed animals at 3 time periods pe. * p < 0.05*

82

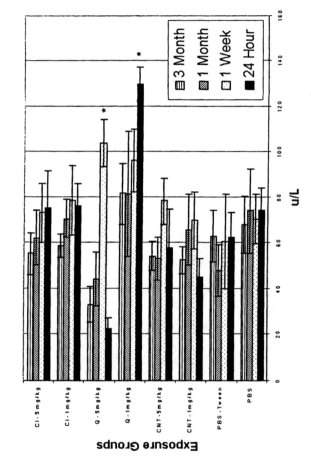

Figure 7. *BAL fluid alkaline phosphatase (ALKP) values for particulate-exposed rats and corresponding controls at 24 hrs, 1 week, 1 month and 3 months postexposure (pe). Significant increases in BALF ALKP vs. controls were measured in the Quartz 1 and 5 mg/kg exposed group at 24 hrs pe and 1 week pe, respectively.* *p < 0.05*

compared to controls (Fig. 8). This alteration in cellular movement represents an adverse alteration in macrophage function and clearance capacity. Exposures to single wall carbon nanotubes or carbonyl iron particles did not result in chemotactic functional deficits of bronchoalveolar lavage-recovered macrophages.

Lung Morphology Studies

Cell Proliferation Results

Lung parenchymal cell proliferation rates (i.e., % immunostained cells taking up BrdU) were measured in particulate-exposed rats and controls at 24 hrs, 1 week, and 1 and 3 months postexposure (pe). Significant enhanced cell proliferation indices were quantified in the high dose quartz-exposed rats at 24 hrs and 1 month postexposure (Fig. 9). No statistically significant increases in airway cell proliferation rates for any of the groups were measured at 24 hrs, 1 week, 1 and 3 months postexposure.

Histopathological Evaluation

Histopathological evaluation of lung tissues revealed that pulmonary exposures to carbonyl iron particles in rats produced no significant adverse effects when compared to PBS-Tween exposed controls.

Pulmonary exposures to quartz particles in rats produced a dose-dependent lung inflammatory response characterized by neutrophils and foamy (lipid-containing) alveolar macrophage accumulation. In addition, early lung fibrosis was evident and progressive (Fig. 10).

Exposures to SWCNT produced a non dose-dependent foreign tissue body as evidenced by the presence of multifocal granulomas, The lesions were initially observed at 1 week postexposure. At 1 month postexposure, a diffuse nonuniform distribution pattern of multifocal macrophage-containing granulomas was present, primarily in the airways. At higher magnification, one could discern the discrete multifocal mononuclear granulomas centered around the carbon nanotube material, most likely in the form of ropes (Fig. 11). There seemed to be little progression of the lesion at 3 months postexposure.

Discussion

This study was designed to assess the pulmonary toxicity of instilled single wall carbon nanotubes (SWCNT) in rats. The lung toxicity of instilled

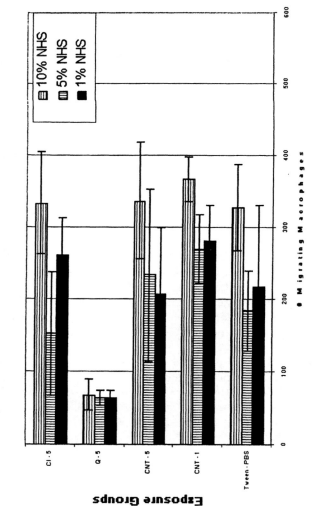

Figure 8. Alveolar macrophage chemotactic capacities in rats exposed to particle-types and corresponding controls at 1 week postexposure. Values given are means ± S.D of chemotactic responses to 1, 5, and 10% NHS chemoattractant. Exposure to quartz particles resulted in functional motility deficits in lavage-recovered alveolar macrophages when compared to controls.

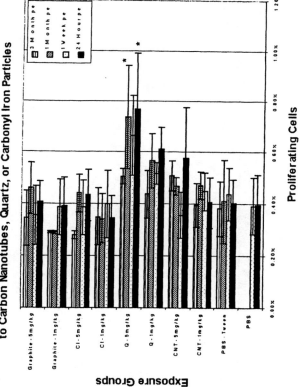

Lung Parenchymal Cell Proliferation Rates of Rats Exposed to Carbon Nanotubes, Quartz, or Carbonyl Iron Particles

*Figure 9. Lung parenchymal cell proliferation rates (BrdU) in particulate-exposed rats and corresponding controls at 24 hrs, 1 week, 1 month and 3 months postexposure (pe). Significant increases in BrdU immunostained cells were measured in the 5 mg/kg Quartz-exposed rats vs. controls at 24 hrs and 1 month pe. * p < 0.05*

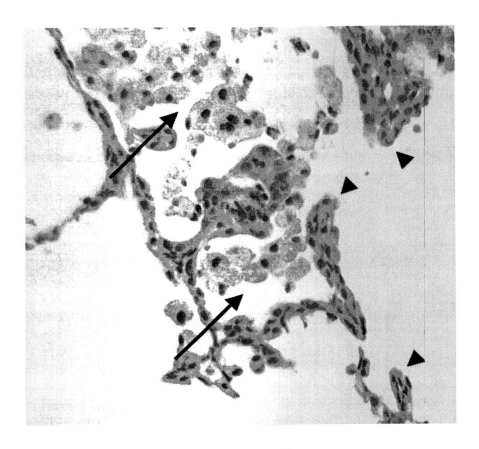

Figure 10. Light micrograph of lung tissue from another rat exposed to quartz particles (5 mg/kg) at 3 months post-instillation exposure. Note the tissue accumulation of foamy multinucleated alveolar macrophages (arrows) within alveolar spaces. The macrophages have migrated to the sites of quartz particle instillation – At the terminal bronchiolar alveolar junctions. The accumulation of lipid-filled macrophages and lack of clearance is a common feature of the progressive nature of silica-induced lung disease. Magnification = x100.

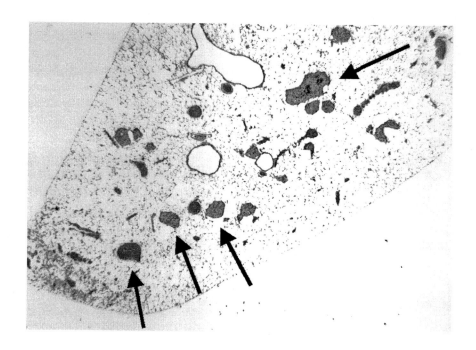

Figure 11. Low magnification micrograph of lung tissue from a rat exposed to carbon nanotubes (1 mg/kg) at 1 month post-instillation. Note the diffuse pattern of granulomatous lesions (arrows). It was interesting to note that few lesions existed in some lobes while other lobes contain several granulomatous lesions – and this was likely due to the nonuniform deposition pattern following carbon nanotube instillation. Magnification = x20.

carbon nanotubes was compared with a positive control particle-type, quartz, as well as a negative control particle-type, carbonyl iron particles. Exposures to SWCNT soot at 5 mg/kg produced mortality in ~15% of the exposed rats. This was due to agglomeration of the major airways by the instillate and not due to inherent toxicity of SWCNT. Results from the bronchoalveolar lavage fluid and cell proliferation evaluations demonstrated that instillation exposures to the positive control quartz particles, particularly at the higher dose, produced persistent adverse effects in lung inflammatory response, cytotoxic effects, and lung cell parenchymal cell proliferation indices. In contrast, SWCNT exposures produced a transient inflammatory response at 24 hours postexposure, related primarily to the clogging of airways and consequent injury by the instillate. Instillation exposures to carbonyl iron particles produced only short-term and reversible lung inflammation, related to the effects of the instillation procedure. Histopathological evaluations revealed that exposures to quartz particles produced pulmonary inflammation, foamy macrophage accumulation and tissue thickening (i.e., fibrosis). Exposures to SWCNT produced multifocal granulomas, evidence of a foreign tissue reaction, and these lesions were generally characterized by a bolus of carbon nanotubes, surrounded concentrically by a macrophage-like granulomatous responses. The finding of granulomas, in the absence of adverse effects measured by pulmonary biomarkers and cell proliferation indices is very surprising, and does not appear to follow the normal inflammogenic/fibrotic pattern produced by fibrogenic dusts, such as quartz, asbestos, and silicon carbide whiskers. Exposure to those dusts produces progressive lesions which can be monitored using bronchoalveolar lavage analyses and additional pulmonary biomarkers of cell injury, inflammation, and fibrosis.

Lam and coworkers (8) assessed the pulmonary toxicity by intratracheal instillation of three nanotube products in mice. These materials were manufactured by different methods and contained different metal catalysts. Accordingly, metal analysis showed that the HIPCO-prepared nanotubes (NT) contained 27% (w/w) iron in the raw form, and 2% iron after purification. Carbolex electric-arc product contained 26% nickel and 5% yttrium. Groups of male B6C3F1 mice each were intratracheally instilled with 0, 0.1, or 0.5 mg of NT suspended in mouse serum. Additional mice were exposed to carbon black and quartz particles. The investigators reported that carbon black particle exposure produced minor pulmonary effects and high-dose quartz produced moderate inflammation in the lung. Moreover, Lam and colleagues reported that all of the NT products, regardless of the type or amount of metal, induced epithelioid granulomas in the centrilobular alveolar septa and, in some cases, interstitial inflammation in the animals of the 7-day postexposure groups. These lesions persisted in the 90-day postexposure groups. The granulomas in NT-treated mice were characterized by the authors as aggregates of macrophages laden with black NT particles. Lam et al. concluded that, if single wall nanotubes reach the lung, they can be more toxic than quartz.

To summarize, instillation of SWCNT in the lungs of rats produced a non dose-dependent a foreign tissue body reaction. The distribution of the multifocal granuloma lesions was nonuniform. There appeared to be no dose response relationship, and a possible regression of lesions occurring from the 1-month to 3-month postexposure periods was noted. The development of these lesions is inconsistent with pulmonary effects observed with fibrogenic dusts, which generally occur at bronchoalveolar junctions. It was interesting to note that the pulmonary biomarkers measured in this study were not predictive of this granulomatous lesion. In addition, two occupational exposure assessment studies of carbon nanotube operations have been recently reported. Both studies have reported very low aerosol exposure levels of respirable-sized carbon nanotubes, ranging from not detectable to < 0.1 mg/m^3 (*9,10,11*). These findings give support to the hypothesis that, due to their electrostatic attraction and proclivity to agglomerate into nanorope structures, aerosol exposures at the workplace to respirable-sized carbon nanotubes are extremely low. Therefore, the study findings of multifocal granulomas that we have reported herein may not have physiological relevance. Thus, to evaluate the true potential risks associated with inhaling single wall carbon nanotubes, the pulmonary effects of SWCNT exposure must be evaluated by conducting an inhalation toxicity study with SWCNT in rats.

Acknowledgments

This study was supported by DuPont Central Research and Development.

Denise Hoban, Elizabeth Wilkinson and Rachel Cushwa conducted the BAL fluid biomarker assessments. Carolyn Lloyd, Lisa Lewis, John Barr prepared lung tissue sections and conducted the BrdU cell proliferation staining methods. Dr. Steven R. Frame provided histopathological evaluations of lung tissues. Don Hildabrandt provided animal resource care.

References

1. Driscoll, KE.; Carter, JM.; Howard, BW.; Hassenbein, DG.; Pepelko, W.; Baggs, RB.; Oberdoerster, G. Pulmonary inflammatory, chemokine, and mutagenic responses in rats after subchronic inhalation of carbon black. *Toxicol. Appl. Pharmacol.* **1996** *136*, 372-80.
2. Heinrich, U.; Peters, L.; Creutzenberg, O.; Dasenbrock, C.; Hoymann, HG. Inhalation exposure of rats to tar/pitch condensation aerosol or carbon black

alone or in combination with irritant gases. *Toxic Carcinog. Eff. Solid Part. Respir. Tract* **1994** pp. 433-44.

3. Nikula, KJ.; Snipes, MB.; Barr, EB.; Griffith, WC.; Henderson, RF.; Mauderly, JL. Comparative pulmonary toxicities and carcinogenicities of chronically inhaled diesel exhaust and carbon black in F344 rats. *Fundam. Appl. Toxicol.* **1995** *25*, 80-94.

4. Rinzler, AG.; Liu, J.; Dai, H.; Nikolaev, P.; Huffman, CB.; Rodriguez-Macias, FJ.; Boul, PJ.; Lu, AH.; Heymann, D.; Colbert, DT.; Lee, RS.; Fischer, JE.; Rao, AM.; Eklund, PC.; Smalley, RE. Large-scale purification of single-wall carbon nanotubes. Process, product, and characterization. *Applied Physics A: Materials Science & Processing* **1998** *A67(1)*, 29-37.

5. Warheit, DB.; Carakostas, MC.; Hartsky, MA.; Hansen JF. Development of a short-term inhalation bioassay to assess pulmonary toxicity of inhaled particles: Comparisons of pulmonary responses to carbonyl iron and silica. *Toxicol. Appl. Pharmacol.* **1991** *107*, 350-368.

6.Warheit, DB.; Hansen, JF.; Yuen, IS.; Kelly, DP.; Snajdr, S.; Hartsky, MA. Inhalation of high concentrations of low toxicity dusts in rats results in pulmonary and macrophage clearance impairments. *Toxicol. Appl. Pharmacol.* **1997** *145*, 10-22.

7. Warheit, DB.; Hill, LH.; George, G.; Brody, AR. Time course of chemotactic factor generation and the corresponding macrophage response to asbestos inhalation. *Amer. Rev. Respir. Dis.* **1986** *134*, 128-133.

8. Lam, C.; James, JT.; McCluskey, R.; Hunter RL. Pulmonary toxicity of carbon nanotubes in mice 7 and 90 days after intratracheal instillation. The Toxicologist **2003** *Vol 72, number S-1*, p 44.

9. Baron, PA.; Maynard, AD.; Foley, M. Evaluation of aerosol release during the handling of unrefined single walled carbon nanotube material. NIOSH Report. NIOSH DART-02-191, December 2002.

10. Joseph, G. Industrial hygiene air monitoring report. DuPont Co. internal report, October, 2002.

11. Maynard, AD.; Baron, PA.; Foley, M.; Shvedova, AA.; Kisin, ER.; Catsranova V. Exposure to carbon nanotube material I: Aerosol release during the handling of unrefined single walled carbon nanotube material. (Submitted for publication).

Environmental Implications: Nanoparticle Geochemistry in Water and Air

Editor
Alexandra Navrotsky

Chapter 10

Environmental Implications: Nanoparticle Geochemistry in Water and Air

Nanoparticles and the Environment

Alexandra Navrotsky

Thermochemistry Facility and NEAT ORU, University of California at Davis, Davis, CA 95616

Nanoscale phenomena are unique and important to environmental science and geoscience. Recent work shows that many outcomes and rates of geochemical processes are governed by phenomena at the nanometer scale, often in small particles. Surface interactions exert a disproportionate influence on the structure, chemistry, and movement of these nanomaterials in the Earth. In many cases, traditional distinctions between solutes, colloids and solids become vague.

Nanogeoscience is broadly defined to include the study of materials and processes at the nanoscale in their role in geologic processes on the Earth and other planets. Because many nanoscale phenomena are concentrated near the Earth's surface, these phenomena are of crucial importance to humans. Because processes are intrinsically molecular at the nanoscale, there is an immediate synergy and a diffuse boundary between nanogeoscience and the fields of chemistry, physics, and materials science. Furthermore, geoscientists increasingly recognize the major role played by microorganisms in geologic phenomena. There is an equally fuzzy boundary between nanogeoscience and the life sciences in this realm because microbial processes often proceed by manipulating surfaces and nanoparticles. Furthermore, nanoparticles in polluted air interact directly with the human respiratory system and may be a major factor in lung disease. Nanoparticles in water both transport and sequester metals and organic matter, leading to changes in bioavailability of nutrients and pollutants. Thus *nanoenvironmental science* focuses on the role of natural and anthropogenic nanoparticles in improving or degrading environmental quality.

Nanoenvironmental science addresses a number of critical environmental issues: the transport of metals and organics in the near-surface environment; global geochemical and climate cycles (including the carbon cycle and global climate change); soil and water quality and sustainable agriculture; atmospheric particle transport, ice nucleation and rainfall; and biological effects in organisms from bacteria to humans. The application of nanoscience and nanotechnology to the environment also addresses national needs: environmental safety, national security, and human health; mining, minerals, oil, and gas; the disposal of chemical and nuclear waste, environmentally friendly manufacturing and new geomimetic materials; and agriculture and food.

Geoscientists have unique skills in aqueous and solid state chemistry, particularly in complex multicomponent systems. They study phenomena occurring on a vast range of scales both in space (nanometers to thousands of kilometers) and time (nanosecond to billions of years). Such a multiscale approach to scientific problems is necessary in monitoring and minimizing the effects of pollution, evaluating the toxicity of materials, and ensuring the safety of water and food supplies.

Much of the chemistry in the atmosphere, on the Earth's surface, and at shallow depths occurs at disequilibrium and minerals form and dissolve in immense solubility gradients. These gradients are caused by bacterial metabolism, by the transition from oxygen-rich to anoxic environments, by the large temperature and pressure gradients intrinsic to the atmosphere, oceans, and shallow Earth. Pollution itself causes sharp gradients in pH and chemical concentration. The initial precipitates that form in these settings can consist of only a few tens of atoms; they are thermodynamically metastable yet some may exist for thousands to millions of years. These very fine-grained materials, or *nanoparticles*, are not only very small but they are also very common. Though they may represent only a small fraction of the *mass* of material in the Earth, they represent a large *number* fraction of the particles in atmospheric and aqueous environments, and nanoparticles are responsible for most of the *surface area* (at solid-water, solid-air, and solid-solid interfaces) of Earth materials. Because chemical reactivity is much greater at surfaces and interfaces than in the bulk, most of the chemical reactions in both natural and laboratory systems disproportionately involve nanoparticles.

Their large surface-to-volume ratio ensures that surface forces exert considerable influence over the chemistry and structure of nanoparticles and nanomaterials in general, to the point that they exhibit properties that are distinct from those of the macroscopic solid. For particles in the size range of roughly 0.5 to 500 nm, distinctions among solutes, molecular clusters, macromolecules, and colloids are vague. Also vague is the distinction between amorphous, disordered, and crystalline solids.

Fine grained oxides precipitated from aqueous solution often crystallize in structures different from those of coarsely crystalline materials: γ-alumina instead of corundum, anatase and brookite instead of rutile, maghemite instead of hematite, and a host of complex hydrous iron oxyhydroxides. It has been suspected that differences in surface energy stabilize, as nanoparticles, polymorphs that are metastable in the bulk. High temperature oxide melt solution calorimetry directly confirmed this enthalpy crossover for γ- versus α-alumina (1) and for anatase and brookite relative to rutile (2). Relative to bulk rutile, bulk brookite is 0.71 ± 0.38 kJ/mol and bulk anatase is 2.61 ± 0.41 kJ/mol higher in enthalpy. The surface enthalpies of rutile, brookite, and anatase are 2.2 ± 0.2 J/m^2, 1.0 ± 0.2 J/m^2, and 0.4 ± 0.1 J/m^2 respectively. The closely balanced energetics directly confirms the crossover in stability of nanophase TiO$_2$ polymorphs inferred by Zhang and Banfield (3). The heavy lines in Figure 1 show the energetically stable phases as a function of surface area. This example shows that new phenomena (in this case, stabilization of polymorphs with quite different properties from those of coarse-grained crystals of rutile) can occur at the nanoscale. Such different properties can affect pollutant adsorption and transport, photocatalysis, toxicity and other processes of environmental importance.

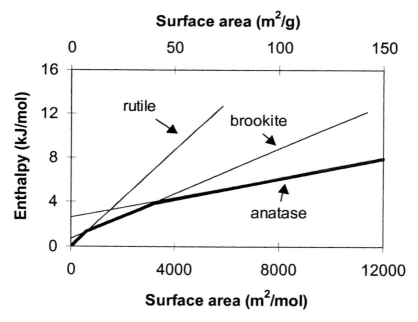

Figure 1. Enthalpy, relative to bulk rutile, of nanophase titania polymorphs, measured by high temperature oxide melt calorimetry

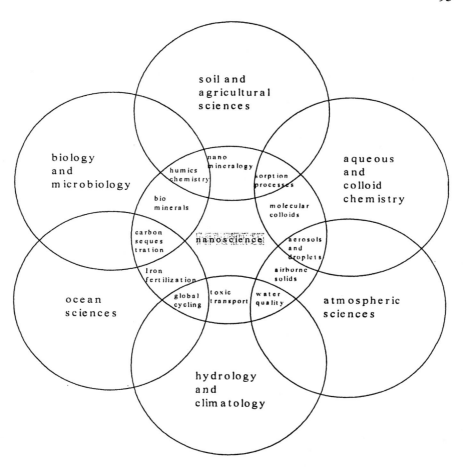

Figure 2. Venn diagram showing overlaps of traditional environmental sciences with nanoscience. We suggest that all of such overlaps define nanoGeoscience.

In a more general sense, nanogeoscience and nanoenvironmental science lie at the center of, and provide fundamental understanding for, a number of environmental applications, as shown in the Venn diagram in Figure 2. The application of nanomaterials and nanotechnology to the detection of contaminants (inorganic, organic, or biological) offers additional important links between environmental science and nanoscale science and technology.

References

1. J. M. McHale, A. Auroux, A. J. Perrotta, and A. Navrotsky (1997) Surface energies and thermodynamic phase stability in nanocrystalline aluminas, Science 277, 788-791
2. M.R. Ranade, A. Navrotsky, H.Z. Zhang, J.F. Banfield, S.H. Elder, A. Zaban, P.H. Borse, S.K. Kulkarni, G.S. Doran, and H.J. Whitfield (2002) Energetics of nanocrystalline TiO_2 Proc. Nat. Acad. Sci. 99, suppl 2, 6476-6481
3. H. Zhang and J.F. Banfield (1998) Thermodynamic analysis of phase stability of nanocrystalline titania. J. Mater. Chem. 8, 2073-2076

Chapter 11

Nanoscale Heavy Metal Phases on Atmospheric and Groundwater Colloids

Satoshi Utsunomiya[1], Kathy Traexler[1], LuMin Wang[1], and Rodney C. Ewing[1,2]

[1]**Nuclear Engineering and Radiological Science, University of Michigan, 2355 Bonisteel Boulevard, Ann Arbor, MI 48109–2104**
[2]**Geological Sciences, University of Michigan, 2534 C. C. Little Building, 425 East University, Ann Arbor, MI 48109 (utu@umich.edu)**

Introduction

Nano-particles are ubiquitous in the ambient environments; atmosphere and groundwater. The effects of fine- to ultra-fine particles in atmosphere on human being (*1,2*) and global climate have been increasing the importance recently (*3*). In groundwater, fine-grain particles (<10um), colloid, has been an issue for transporting toxic and radioactive elements to a far field (*4*). Especially the recent research suggested that the Pu transportation to the far field by colloids at Nevada Test Site (*5*).

Although particles >1 um are characterized well by some analytical methods using X-ray and ion microprobe, a transmission electron microscopy (TEM) is the best technique to characterized nano-particles (< 1um) in environments. However there is still a challenge to investigate trace element phases, which frequently occur in dispersed nano-particles, because the contrast in TEM is usually not correlated to the compositional variety. The recently developed, scanning TEM with high-angle annular detector (HAADF-STEM), allows a compositional imaging in near atomic level, in which the contrast is related to atomic mass and the specimen thickness (*6*). In this study, the atmospheric particulates from Detroit urban area and the colloids from Nevada Test Site.

97

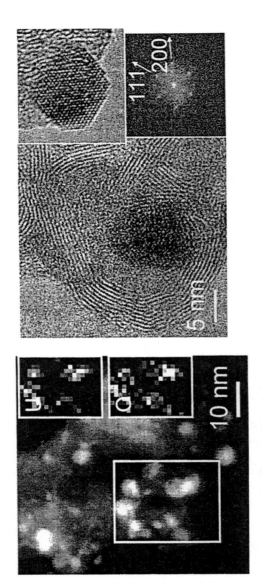

Figure 1. The occurrence of uranium by HAADF-STEM and HRTEM image of U-bearing nanoparticles encapsulated in a "cage" of carbonaceous fulleroid. The particles were subsequently identified as uraninite, UO_{2+x}.

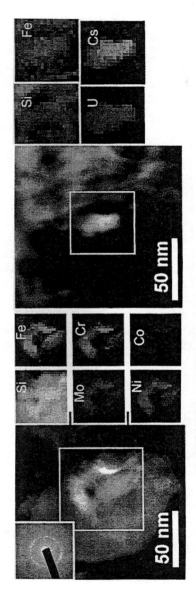

Figure 2, Ground water colloids from Nevada test site. Left picture is a HAADF-STEM image with elemental map of Co-bearing colloids. Inset is the SAED pattern. Right picture shows a Cs-bearing particle.

Experimental method

The samples were characterized using Field Emission SEM (FE-SEM: Philips XL-30), and high resolution TEM (HRTEM: JEOL 2010F). Nano-scale elemental mapping on nanocrystals containing heavy metals was conducted using HAADF-STEM with EDX mapping system (Emispec, ES Vision ver. 4.0) (7). TEM specimens were prepared by dispersing the samples from the filters onto holey carbon grids. Before STEM analysis, the TEM specimen holder was cleaned with plasma (Fischione Model C1020) to minimize a contamination. A drift correction acquisition system was used for the STEM-EDX mapping.

Results and discussion

In Detroit aerosol sample, uranium nanocrystals were detected by HAADF-STEM and subsequently the phase was identified as uraninite, UO_2. In addition, the uraninite nanocrystals were often encapsulated by well-crystalline carbon forming "cage-like" structure. The structure possibly retards the uraninite from alteration in the ambient atmosphere (8). In the same sample, trace elements; Pb, As, La, Ce, Sr, Ti, Zn, Cr, Se, Sn, Y, Zr, Au, and Ag, were detected and the elemental distributions were directly mapped in nano-scale. The structure of the particles bearing Pb, Sr, Ti, Zn, and Au were successfully determined by the diffraction pattern. The size of most particles bearing trace elements detected in this study was distributed within the range of 0.01–1.0 μm, which has the longest residence time, ~100 days, in any particle sizes.

In the groundwater sample from Nevada Test Site, the three important potentially radionuclide elements, Co, Cs, and Eu were identified. Cobalt occurred in awaruite (Ni_2Fe) structure with Mo and Cr. The Cs-bearing phase consisted of Cs, O and U possibly forming a Cs-uranate. An Eu-bearing phase was monazite. The three elements were present, not as absorbed species, but as incorporated in the minerals structure.

References

1. Pope, C. A., III; Burnett, R. T.; Thun, M. J.; Galle, E. E.; Krewski, D., Ito, K.; Thurston. G. D. *J. Am. Med. Assoc.* **2002,** *287*, 1132–1142.
2. Oberdörster, G. *Phil. Trans. R. Soc. Lond. A* **2000,** *358*, 2719–2740.
3. Andreae, M. O. *Nature* **1996,** *380*, 389–390.

4. McCarthy, J. F.; Zachara, J. M. *Environ. Sci. Technol.* **1989**, *23*, 496–502.
5. Kersting A, B. et al. *Nature* **1999**, *397*, 56–59.
6. Pennycook, S. J.; Jesson, D. E. *Phys. Rev. Lett.* **1990**, *64*, 938–941.
7. Utsunomiya, S.; Ewing, R. C. *Environ. Sci. Technol.* **2003**, *37*, 786–791.
8. Utsunomiya, S.; Jensen, K. A.; Keeler, G. J.; Ewing, R. C. *Environ. Sci. Technol.* **003**, *36*, 4943–4947.

Chapter 12

Recent Insights into the Formation and Chemical Composition of Atmospheric Nanoparticles from the Aerosol Nucleation and Realtime Characterization Experiment

J. N. Smith[1,*], K. F. Moore[2], F. L. Eisele[1,3], A. K. Ghimire[4], H. Sakurai[4], and P. H. McMurry[4]

[1]Atmospheric Chemistry Division, National Center for Atmospheric Research, Boulder, CO 80305
[2]Advanced Study Program, National Center for Atmospheric Research, Boulder, CO 80305
[3]Earth and Atmospheric Sciences, Georgia Institute of Technology, 311 Ferst Drive, Atlanta, GA 30332
[4]Mechanical Engineering Department, University of Minnesota, 111 Church Street SE, Minneapolis, MN 55455

Introduction

Recent atmospheric observations have provided important insights into conditions that lead to new particle formation from gas phase precursors, or nucleation. Nucleation sometimes follows regular diurnal patterns, with peak particle production rates occurring in midday when solar radiation is most intense1. Nucleation can also occur in response to atmospheric perturbations, such as the removal of pre-existing aerosol by cloud processing2 or the addition of gas phase reactants from a surface source3, but almost always occurs during daylight hours. Nucleation has been observed on mountains4, in the boreal forests of Finland5, in polluted urban centers6, and in the marine boundary layer7. Impactor measurements of freshly nucleated aerosol in a Finnish boreal forest suggest that these particles are enriched with dimethyl amine8, while other measurements in a variety of locations suggest an important role for sulfuric acid vapor9. Another interesting observation is that the appearance of new particles is sometimes preceded by the continuous growth of ions from small sizes (mobilities between 3.2 and 0.32 cm2V-1s-1 corresponding to diameters between 0.4 and 2.5 nm) up to large sizes (mobilities as low as 0.01 cm2V-1s-1 corresponding to a 50 nm diameter particle)10.

While nucleation under a wide range of circumstances is now well documented, we do not yet have accurate models for predicting nucleation rates. Observed rates are occasionally consistent with the predictions of the binary theory of sulfuric acid and water11, but rates of particle formation are often orders of magnitude higher than can be explained by binary theory12. A ternary process involving sulfuric acid, water, and ammonia13 (or possible an organic compound such as an amine) may explain these high rates of particle production, but additional atmospheric observations are needed to verify this. Theories exist that describe nucleation on ion centers (ion induced nucleation)14, yet there have been no measurements of the composition of atmospheric ions during nucleation events to validate these.

The Aerosol Nucleation and Realtime Characterization Experiment (ANARChE), was a multi-investigator study that focused primarily on understanding the nucleation and growth of atmospheric aerosols. The study took place in midtown Atlanta, GA, from July 22 to September 4, 2002. This site was chosen because three years of continuous measurements at this site have shown that nucleation occurs regularly in August6. ANARChE brought together, for the first time, gas phase measurements of important precursors to new particle formation with several unique instruments for studying the composition of nanoparticles and ambient ions. We present below a review of the most important observations from this study.

Methods

Table 1 shows a summary of the measurements that were performed during ANARChE. Among these, two recently developed techniques are capable of characterizing the chemical composition of freshly nucleated atmospheric nanoparticles. The first instrument, a Nano Tandem Differential Mobility Analyzer (Nano TDMA), characterizes the hygroscopicity and volatility of 3-10 nm diameter atmospheric aerosol15. It performs this measurement by exposing size-classified ambient aerosol to 90% relative humidity (RH) at ambient temperature or to a high temperature (100 °C). A second size classifier is used to observe the increase or decrease in aerosol size compared to its original size, resulting in a value called a "growth factor" which is defined as the ratio of the output to input aerosol diameter. Insights into the chemical composition of the aerosol can be made from these measurements, e.g., sulfate aerosol that is totally or partially neutralized by ammonia will appear as non-volatile (growth factor ~1 when heated to 100 °C) but quite hygroscopic (growth factor ~1.4 when exposed to 90% RH). The second recently developed instrument that was deployed in this study, the Thermal Desorption Chemical Ionization Mass Spectrometer (TDCIMS), can directly measure the chemical composition of 4 – 20 nm diameter aerosol16,17. In the TDCIMS, ambient nanoparticles are charged and then collected on a metal filament for a period of ca. 10 minutes. The filament is moved into the ion source of a chemical ionization mass spectrometer (CIMS), where it is heated to volatilize the particles. The evaporated molecules are then chemically ionized and mass analyzed according to typical CIMS procedures. The most common measurements made during ANARChE with the TDCIMS were back-to-back measurements of ambient aerosol and laboratory-generated particles. Since both types of particles were characterized using identical measurement conditions, the resulting spectra could be directly compared to provide quantitative information on the composition of ambient aerosol.

In addition to the two instruments mentioned above, one more from Table 1 will be described. The flow opposed drift tube tandem mass spectrometer is an instrument that can measure the composition of ambient ions with high sensitivity (a few ions per cm3) and over a wide range of molecular weights (up to 700 amu)18. Clusters at the upper limit of this mass range should be nearly 1.5 nm in diameter. The instrument directly samples ambient ions at high flow (2000 cm s-1). The ions are then concentrated and extracted from the main flow with strong electric fields, and injected into a triple quadrupole mass spectrometer. The ambient ion spectra can then be examined for evidence of cluster growth that would suggest ion induced nucleation. Under the conditions of high levels of sulfuric acid vapor found in Atlanta, the most likely candidate for study would be clusters of $HSO_4^-\cdot(H_2SO_4)_n$. If there were a series of $HSO_4^-\cdot(H_2SO_4)_n$ peaks extending continuously to large values of n, it would suggest a formation process that is consistent with ion induced nucleation.

Results and Discussion

Aerosol Composition Measurements

During the study, eight intense nucleation events were observed, distinguished by concentrations of sub-10 nm diameter aerosol of up to 106 particles cm-1. One example is shown in figure 1a where bursts of small particles are observed between 13:30 and 16:30 EDT on August 25, 2002. Figure 1b shows the results of concurrent Nano TDMA measurements of aerosol volatility for 12 nm diameter particles. The measurements show that the aerosol that formed as a result of the nucleation events were less volatile than those at other times that day. The results of concurrent TDCIMS measurements for a narrow distribution (< 4 nm FWHM) of particles centered at 15 nm in diameter are shown in figure 1c. The data are expressed as the ratio of the ambient aerosol sulfate to sulfate from the same mass of pure ammonium sulfate aerosol, a value we call the "$(NH_4)_2SO_4$ mass fraction." A value of one suggests that the ambient aerosol may be pure ammonium sulfate. Figure 1c shows that the aerosol that formed as a result of nucleation is consistent with sulfate aerosol that is almost completely neutralized by ammonium. This result that ammonium sulfate seems to be the dominant compound observed during nucleation is consistent with the volatility measurements performed by the Nano TDMA (figure 1b), since as mentioned previously ammonium sulfate aerosol are non-volatile. These results are typical of those obtained from other nucleation events during ANARChE. An example of a measurement of aerosol hygroscopicity is shown in figure 2, where it can be concluded that 10 nm diameter aerosol resulting from nucleation events are more hygroscopic than those at other times of the day. This again is consistent with a completely or partially neutralized ammonium sulfate aerosol. When ammonium was measured and compared to calibration aerosols in the same manner as sulfate in figure 1c we found that the ammonium composition of ambient freshly nucleated aerosols was also consistent with ammonium sulfate aerosol. One important result of TDCIMS measurements performed during the study is that, in spite of our best efforts at searching for other compounds in the aerosols, our measurements showed that the sub-20 nm diameter aerosols were composed almost entirely of ammonium and sulfate. Thus it seems that the preponderance of evidence points to the fact that newly formed aerosol in Atlanta during ANARChE are composed primarily of sulfate that is almost completely neutralized by ammonium.

Ambient Ion Composition Measurements.

An example of negative ion composition measurements performed using the flow opposed drift tube tandem mass spectrometer is shown in figure 3. From past measurements of natural atmospheric ion spectra, we already know that $H_2SO_4^-$ and $HSO_4^- \cdot (H_2SO_4)$ are common daytime ions since H_2SO_4 is generated through a multi-step photochemical process. At night when gas phase H_2SO_4 is largely scavenged by existing aerosol particles the ion spectrum is expected to be dominated by NO_3^- ions and its H_2O and HNO_3 clusters. As seen from figure 3, although total ion concentrations are very low, NO_3^- does in fact typically dominate the ion spectrum at night in the absence of H_2SO_4, and

HSO4-·H2SO4 typically dominates the spectrum during the day. There are also, on occasion, some clusters that appear to include H2SO4 and HNO3, but the larger HSO4-·(H2SO4)2·(NH3)m·(H2O)p peaks for m+p=0,1,2 appear to be very small. There is some evidence that the HSO4-·(H2SO4)2 peak is not quite zero, but it is less than 10% of the HSO4-·H2SO4 peak which is inconsistent with rapid ion growth to the larger clusters expected for ion induced nucleation. In the positive spectra, there do not appear to be any clearly consistent peaks, only a few small variable peaks and a small concentration of many different masses. If enhanced clustering is occurring in the positive spectrum, it must be with a wide variety of different mass compounds. While the ion composition results do not suggest that ion-induced nucleation played a significant role during ANARChE, it is still possible that ions could have led to some enhancement in neutral molecular cluster formation by providing ion-ion recombination products that are more stable than those formed through neutral reactions above. These might include gas phase clusters or molecules containing one or two H2SO4 molecules (which dominated the negative spectrum) plus a high proton affinity (typically alkaline) compound present in the positive ion spectrum.

References

1. Bradbury, N. E.; Meuron, H. J. Terr. Magn. 1938, 43, 231-240.
2. Hegg, D. A.; Radke, L. F.; Hobbs, P. V. J. Geophys. Res.-Atmos. 1990, 95, 13917-13926.
3. McGovern, F. M.; Jennings, S. G.; Oconnor, T. C. Atmos. Environ. 1996, 30, 3891-3902.
4. Shaw, G. E. Atmos. Environ. 1989, 23, 2841-2846.
5. Makela, J. M.; Aalto, P.; Jokinen, V.; Pohja, T.; Nissinen, A.; Palmroth, S.; Markkanen, T.; Seitsonen, K.; Lihavainen, H.; Kulmala, M. Geophys. Res. Lett. 1997, 24, 1219-1222.
6. Woo, K. S.; Chen, D. R.; Pui, D. Y. H.; McMurry, P. H. Aerosol Sci. Technol. 2001, 34, 75-87.
7. Covert, D. S.; Kapustin, V. N.; Quinn, P. K.; Bates, T. S. J. Geophys. Res.-Atmos. 1992, 97, 20581-20589.
8. Makala, J.; Mattila, T.; Hiltunen, V. American Association for Aerosol Research, 1999, Tacoma, WA.
9. F. L.; McMurry, P. H. Philos. Trans. R. Soc. Lond. Ser. B-Biol. Sci. 1997, 352, 191-200.

10. Horrak, U.; Salm, J.; Tammet, H. J. Geophys. Res.-Atmos. 1998, 103, 13909-13915.
11. Weber, R. J.; McMurry, P. H.; Mauldin, R. L.; Tanner, D. J.; Eisele, F. L.; Clarke, A. D.; Kapustin, V. N. Geophys. Res. Lett. 1999, 26, 307-310.
12. Weber, R. J.; Marti, J. J.; McMurry, P. H.; Eisele, F. L.; Tanner, D. J.; Jefferson, A. Chem. Eng. Commun. 1996, 151, 53-64.
13. Korhonen, P.; Kulmala, M.; Laaksonen, A.; Viisanen, Y.; McGraw, R.; Seinfeld, J. H. J. Geophys. Res.-Atmos. 1999, 104, 26349-26353.
14. Yu, F. Q.; Turco, R. P. Geophys. Res. Lett. 2000, 27, 883-886.
15. Sakurai, H., Tobias, H.J., Park, K., Zarling, D., Docherty, K.S.,; Kittelson, D. B., McMurry, P.H. Atmos. Environ. 2003, in press.
16. Voisin, D.; Smith, J. N.; Sakurai, H.; McMurry, P. H.; Eisele, F. L. Aerosol Sci. Technol. 2003, 37, 471 - 475.
17. Smith, J. N.; Moore, K. F.; McMurry, P. H.; Eisele, F. L. Aerosol Sci. Technol. 2003, submitted.
18. Eisele, F. L. J. Geophys. Res. 1988, 93, 716.

Chapter 13

Outer-Sphere Coordination of Uranyl Carbonate by Modified Cyclodextrins: Implications for Speciation and Remediation

Jason R. Telford

Department of Chemistry, University of Iowa, Iowa City, IA 52242–1294

The characterization of two outer-sphere ligands with uranyl carbonate are presented. The ligands are based on a- and b-cyclodextrin architectures, with amine hydrogen bonding groups appended to the 1° hydroxyl ring of the cyclodextrin. Binding affinities for uranyl carbonate are K_a = 253 M^{-1} and 838 M^{-1} for per-[6-(ethylenediamino)]-6-deoxy-a-cyclodextrin (**1**) and per-[6-(amino)]-6-deoxy-bcyclodextrin (**2**) respectively. These ligands are able to complex uranyl carbonate in seawater to an appreciable extent, and influence the speciation of the metal complex.

Introduction and Background.

Nature controls metal-ion reactivity through inner- and outer-sphere coordination chemistry. Inner-sphere coordination is provided through the few amino acids and cofactors capable of binding a metal, and outer-sphere influences are a result of protein secondary and tertiary structure. In all instances characterized to date, proteins recognize small inorganic substrates, such as phosphate or iron complexes, by providing a binding site with a spatial arrangement of hydrogen bonds or hydrophobic patches complementary to their substrate.[1-3] In these natural systems, there is no exchange of ligands (replacement of a water by a protein-based ligand, for example), but rather the inorganic *complex* is recognized *in situ*. We have developed supramolecular ligands which bind small inorganic complexes using the same strategy Nature employs, namely outer-sphere recognition.[4,5] The ability to bind metal guests by outer-sphere coordination has far-reaching implications to applied environmental and industrial remediation. Instead of protein architectures, here we report the use of cyclodextrins (CyD) as the molecular architectures (Figure 1). There is no metal-binding by the CyDs themselves, but our CyD derivatives are modified to present a hydrogen-bonding pocket for a target metal-complex. These ligands take advantage of the unique geometric features of a substrate, with recognition modulated by interaction with the guest topology, van der Waals interactions, and hydrogen bonding. While several research groups target oxoanions such as phosphate, sulphate and nitrate in host-guest systems[6,7], our target for this study is the synthesis and characterization of a family of outer-sphere coordination complexes with $[UO_2(CO_3)_3]^{4-}$-(uranyl carbonate).[8] Outer-sphere coordination (interaction with the metal's pre-existing ligands rather than bonding directly to the metal) is a tunable approach to binding the uranyl ion, and thus provides a more selective approach to environmental uranyl remediation.

There are several methods currently utilized to remediate contaminated sites. Treatment by bioremediation (uptake by microbes) and extraction technologies are currently being used and investigated.[9-11] Another current environmental remediation strategy relies on removing metals, including the transuranic elements, by complexation with powerful chelating agents.[12,13] These chelating ligands form thermodynamically stable complexes, so in order to remove the metal and reuse the ligands, a high thermodynamic cost must be paid. The advantage of our approach is that a pre-existing metal complex can be targeted *in situ*, bypassing the need for costly adjustment of solution conditions to favor a new metal complex and bypassing the high thermodynamic cost of regenerating the ligand.

Environmentally, virtually all soluble uranium is present as uranyl carbonate.[14,15] The carbonates of the *trans*-oxo uranyl complex are in a trigonally symmetric arrangement about the equatorial plane (Figure1).[8,15] The terminal oxo anions of the carbonates are strong hydrogen bond acceptors. These simple structural features- a unique geometry and well-defined hydrogen bond accepting groups- coupled with the pH-sensitive chemistry of uranyl ion- provide a handle by which we can control the chemistry of the uranyl ion in aqueous environments.

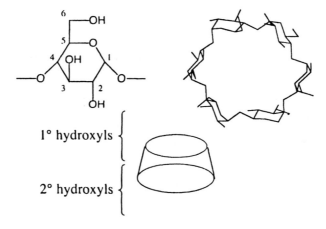

6 —OH

5 —O

4 OH 1

—O O—

3 2

OH

1° hydroxyls

2° hydroxyls

Figure 1. Three views of α-cyclodextrin showing numbering scheme for monomer units (upper left), molecular architecture of α-cyclodextrin (upper right) and cartoon of toroidal shape and hydroxyl disposition (lower). The edge ringed with 1° hydroxyls is referred to as the upper rim.

Materials and Methods.

General.

NMR spectra were obtained using a Bruker 400 MHz instrument for 1H and 13C NMR (100 MHz). Chemical shifts are reported as parts per million (ppm) from tetramethylsilane (TMS). Chemicals and solvents were commercial reagent grade and used without further purification. Water used for titrations was purified by reverse osmosis (MilliQ, Millipore Corp.) to greater than 18 $M\Omega$ resistance. A modification of the literature procedures described previously for the preparation of per-[6-(ωalkylamino)]-6-deoxy-cyclodextrin was employed for the compounds reported here.[16,17]

Experimental.

Preparation of per-6-amino-6-deoxy-cyclodextrins per-[6-(ethylenediamino)]-6-deoxy-a-cyclodextrin (1). A mixture of 1.45 g (0.75 mmol) of per-6-iodo-6-deoxy-a-cyclodextrin in 10 ml of 1,2-diaminoethane was heated at 60 oC for 14-16 h. After evaporation of excess amine, 50 mL of absolute EtOH was added. The precipitate which formed was collected and washed with absolute EtOH to give the product as the HI salt, which was either purified by dialysis (cellulose membrane) with ammonium bicarbonate solution or chromatographed on LH20 eluting with water to afford product. 13C (D2O) d 104.1 (C1), 86.0 (C4), 75.8 (C3), 74.3 (C2), 73.3 (C5), 53.9 (NHCH2), 51.9 (C6), 42.4(CH2NH2), MS (+FAB) M+H+ 1219. per-[6-(amino)]-6-deoxy-b-cyclodextrin (2). Was prepared as previously described.[17] 13C (D2O) d 104.0 (C1), 86.1 (C4), 75.5 (C3), 74.7 (C2), 73.5 (C5), 55.5 (C6) MS(+FAB) M+H+ 1128. Uranyl carbonate was generated *in situ* by dissolving uranium nitrate into a K2CO3/D2O buffer. This solution was passed down a short column (Sephadex, G-10) using carbonate buffer as an eluent. The concentration of the resultant solution was calculated by UV/vis absorbance.

Titrations.

Titrations were repeated at least three times using slightly varied concentrations of ligand. In general, a solution of 10 mM (**1**) or (**2**) was made up in 0.5 mL D2O in an NMR tube. The solution was buffered to pD 8.0 by addition of K2CO3 (100 mM). Small aliquots of uranyl carbonate (0.115 M UO2) in K2CO3 /D2O buffer were added directly to the NMR tube to make up a solution ranging from 0-100 mM. The 1H NMR spectrum was acquired after each addition. A small amount of DMF (**1**) or methanol (**2**) was added as an internal standard to calibrate the spectrum.

The data, which comprise a series of 10-15 spectra were imported into the program HypNMR.[18] Refinement of the observed proton shifts as a function of sequential uranyl carbonate additions yields an observed stability constant consistent with a 1:1 association. An assumption is implicit with the data evaluation, which is that the equilibrium between chemical species within the titration is rapid on the NMR time-scale. The chemical shift for a given nucleus is assumed to be a concentration-weighted average over all of the chemical species in which the nucleus is present. This assumption appears valid for these weakly bound complexes. There is no evidence for slow exchange within the NMR titration experiment.

Results and Conclusions.

Synthesis.

To date, there are few examples of metal complex recognition by synthetic, outer-sphere ligands.[19-22] Outer-sphere binding does not involve making strong covalent bonds to the metal center. Rather, the summation of multiple weak forces such as hydrogen bonds or electrostatic attraction must provide the energetic requirements for complexation. To meet this end, the architecture upon which outer-sphere ligands are constructed plays an important role in preorganizing the binding groups to maximize enthalpic binding as the entropic component of desolvating a charged metal complex is unfavorable.[23,24]

Synthesis of upper rim cyclodextrin derivatives follows outlines of published procedures to give compounds of the general type illustrated in Figure 3.[16,17,25,26] Three pendant arms are shown for clarity, but we have generated cyclodextrin derivatives with 6 or 7 pendant groups. Conversion of the cyclodextrin to the per-6-iodo-cyclodextrin is accomplished by treatment with I_2/triphenylphosphine in DMF. This modification is reported in the literature for β-cyclodextrin, and proceeds well in our laboratory with α- or β-cyclodextrin.

Using this reaction scheme, we have generated the per-6-(1,2-ethylenediamine)-6-deoxy-α-cyclodextrin and per-6-(amino)-6-deoxy-β-cyclodextrin. Per-6-(amino)-6-deoxy-β-cyclodextrin (Figure 3) is synthesized by treating the parent iodo compound with sodium azide in DMF. Precipitation with acetone affords the per-azidoβ–cyclodextrin in quantitative yield. The azide is reduced to the amine by treatment with triphenylphosphine and

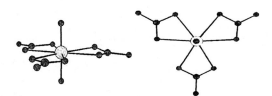

Figure 2. Two views of the uranyl carbonate anion. Side view (left) and top (right)

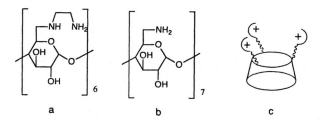

Figure 3. a) per-[6-(ethylenediamino)]-6-deoxy-α-cyclodextrin (1), b) per-[6-(amino)]-6-deoxy-β-cyclodextrin (2), c) cartoon diagram of ligands, with only three hydrogen bonding groups shown for clarity.

ammonium hydroxide at room temperature. Precipitation from 9 ethanol yields per-6-amino-β–cyclodextrin.

The per-6-(1,2-ethylenediamine)-6-deoxy-α-cyclodextrin (Figure 3) is synthesized in high yield from the iodo compound by reacting with neat ethylene diamine. Following isolation, this highly water-soluble compound was fully characterized. The mass spectrum of the compound shows the appropriate M+H peak, as well as the +1, 2, 4, and 6 counterions. The NMR spectra ($_1$H and $_{13}$C) are relatively simple at low pH, indicating the amine substituents do not disrupt the 6-fold symmetry of the backbone (Figure 4). At high pH, all peaks are broader, as is common in reported work with βcyclodextrin. [17]

Uranyl ion coordination and thermodynamics.

The U(VI) ion is diamagnetic, which allows straightforward NMR experiments on the ligands and on the metal complexes. Thermodynamic evaluation of the binding affinity between per-6-(1,2-ethylenediamine)- 6-deoxy-α-cyclodextrin and uranyl carbonate was examined by $_1$H NMR.[27] Titrations of this ligand with uranyl carbonate (0-2 equivalents, 10 mM [ligand]) show a metalconcentration dependent upfield shift of ~0.1 ppm of the ethylene-diamine CH$_2$ protons which suggests an interaction of the (terminal) 1° amines with either the uranyl or the carbonate ligands. A direct interaction of the amines with the uranyl ion is highly unlikely based on the extremely low association constants of amines with uranyl.[28,29] A simple pH or ionic strength effect can be ruled out, as the titration was carried out under Figure 4. $_1$H NMR of per-6-ethylenediamine-6-deoxy-α-cyclodextrin.

The single C1 proton signal indicates a C$_6$symmetry for the molecule. 10

conditions of constant pH and ionic strength. Additionally, small but definite shifts in the C1 and C4 protons as well as other sugar protons are observed (Figure 5). The shift in the C1, C4 and methylene proton peaks can be fit to a binding isotherm to give K_a = 253 M-$_1$. The complexation-induced chemical shift of the C4 protons, as well as the other sugar-based protons, suggests that binding is at least partially in the cavity of the torus.[30]

The data fit well to a 1:1 stoichiometry and there is no evidence for binding stoichiometries greater than 1:1. Similar behavior is seen in the titration of per-6-(amino)-6-deoxy-β-cyclodextrin. Again, clear shifts in the C1, C4, C5 and C6 protons can fit to a binding isotherm. In the case of this slightly larger architecture, a fit of the data gives a calculated stability constant, K_{assoc}, of 838 M-$_1$. chemical shifts for nucleus 2 6 10

Point number 5.070 5.080 5.090 5.100 5.110

Chemical shift unweighted error in chemical shift. (Weighted rms = 0.002) 2 6 10 -0.004 -0.002 0.000 0.002 0.004

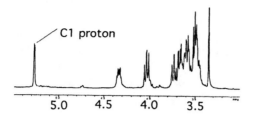

Figure 4. ¹H NMR of per-6-ethylenediamine-6-deoxy-α-cyclodextrin. The single C1 proton signal indicates a C₆ symmetry for the molecule.

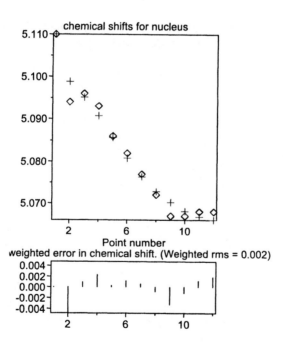

Figure 5. Observed (◇) and calculated (+) chemical shift change Δδ in ¹H NMR of C1 proton (right) of per-6-ethylenediamine-6-deoxy-α-cyclodextrin as function of added metal.[27] Titration conditions are given in text.

This increase in binding affinity is probably due in large part to the increased size of βcyclodextrin. While α– and β–cyclodextrin have similar distances from 1° to 1° hydroxyl (maximum of 11.5 – 12 Å across the smaller end of the torus) because of the rotation of the C6 methylene groups, the internal cavity of β-cyclodextrin is considerably larger and likely is better able to accommodate the large uranyl carbonate ion. We hypothesize that the uranyl complex is bound as a unit by the amino-CyD hosts, based on the low affinity of nitrogen bases for U(VI) ions and the large cavity of the hosts. It is not likely that the uranyl ion is bound in the hydrophobic internal cavity of the CyD, but rather hydrogen bonded to the amino groups at the 1° edge of the torus.

Species Distribution Studies.

The distribution of species in aqueous solutions of uranyl ion and carbonate is complex, and highly pH dependant. However, since we know the protonation behavior of the aquated ion, and the formation constants with small inorganic ions such as sulfate, chloride, phosphate and carbonate, we can predict the speciation as a function of concentration, pH, exogenous ligands and other factors.[31] This predictive ability is a powerful tool in the design of efficatious ligands. Figure 6 shows a species distribution of uranyl ion in seawater.

Using the calculated stability constants of one of our ligands (per-6-(1,2-ethylenediamino)-α-cyclodextrin) with uranyl carbonate, we can model the perturbation in seawater speciation if a ligand is present. With an association constant of K= 253 M-1 it can be seen that the addition of ligand strongly stabilizes the tris-carbonato complex over a wider pH range (Figure 7). With a binding constant of 838 M-1, a model of the uranyl-carbonate-(2) system predicts that the carbonate complex will be strongly bound throughout a wide pH range (Figure 8). The protonation states of (1) and (2) were not included in the model. Undoubtably, the ligands (1) and (2) will be deprotonated above pH 10, and their binding affinity substantially reduced from the predicted values at that pH. However, it is likely that the amine H-bond donors will remain predominately protonated at near-neutral pH.

This data shows that metal ion-complex chemistry can be controlled by outer-sphere coordination. By careful design of binding groups and an architectural scaffold we have demonstrated that cyclodextrin-based ligands, such as those described within this paper will form outer-sphere complexes with uranyl carbonate within the pH and concentration range of aqueous environmental conditions such as seawater (pH 8-8.3). Although not all uranyl carbonate is complexed by (1) in this model, the development of ligands with higher binding constants (even one order of magnitude) will result in essentially complete encapsulation of the target substrate. The magnitude of interaction

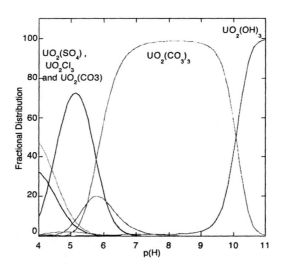

Figure 6. Species distribution of uranyl ion in seawater. The concentration of uranyl is 10^{-4} M. The distribution of uranyl species is relatively insensitive between 10^{-8} and 10^{-4}, but hydroxo complexes dominate at higher concentrations. Carbonate concentration is 1 mM. Other included ions (sulfate, chloride) do not make substantial contributions. The main species below pH 4 is free uranyl (aquo species), from pH 4-7 it is the bis-carbonate, and from pH 7-10, $UO_2(CO_3)_3^{4-}$. Above pH 10, hydroxo complexes dominate.

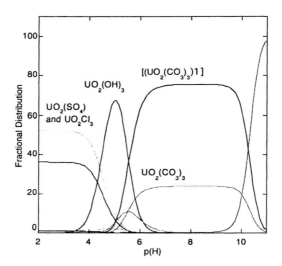

Figure 7. Calculated species distribution of uranyl ion in seawater. The concentrations of species are the same as Figure 6 with the addition of 10 mM ligand (1), $K_a=253$ M^{-1}. The biscarbonate is now the main species from pH 4-5.5 and the triscarbonate dominates from 6-10.5. Although only about 70% of uranyl carbonate is bound, the pH distribution of the uranyl-carbonate species is perturbed.

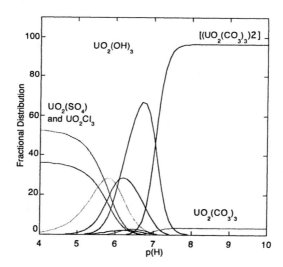

Figure 8. Calculated species distribution of uranyl ion water in seawater. The concentrations of species are the same as Figure 6 with the addition of 10 mM ligand (2), $K_a=838\ M^1$. The biscarbonate is now the main species from pH 4-5.5 and the triscarbonate dominates from 6-10.5. At this level of complexation, the calculated pH distribution of the uranyl-carbonate species is strongly perturbed.

suggests that these ligands, or similar topology, attached to a solid support can act as a reusable resin to remediate uranyl carbonate from aqueous environmental waste.

References.

1. Buchanan, S. K.; Smith, B. S.; Venkatramani, L.; Xia, D.; Esser, L.; Palnitkar, M.;
Chakraborty, R.; van der Helm, D.; Deisenhofer, J. *Nat. Struct. Biol.* **1999**, *6*, 56-63.
2. Jacobson, B. L.; Quiocho, F. A. *J. Mol. Biol.* **1988**, *204*, 783-787.
3. Ledvina, P. S.; Tsai, A. I.; Wang, Z.; Koehl, E.; Quiocho, F. A. *Prot. Sci.* **1998**, *7*, 2550.
4. Zamaraev, K. *New. J. Chem.* **1994**, *18*, 3-18.
5. Loeb, S. J. In *Comprehensive Supramolecular Chemisitry*; Atwood, J. L., Davies, J.
E. D., Macnicol, D. D., Vögtle, F., Volume Ed. Gokel, G. W., Eds.; Elsevier Science Ltd.: Oxford, 1996; Vol. 1, pp 733-754.
6. Hossain, M. A.; Llinares, J. M.; Powell, D.; Bowman-James, K. *Inorg. Chem.* **2001**, *40*, 2936-2937.
7. Gerasimchuk, O. A.; Mason, S.; Llinares, J. M.; Song, M. P.; Alcock, N. W.; Bowman-James, K. *Inorg. Chem.* **2000**, *39*, 1371-1375.
8. Coda, A.; Giusta, A. D.; Tazzoli, V. *Acta Cryst* **1981**, *B37*, 1496-1500.
9. Neunhauserer, C.; Berreck, M.; Insam, H. *Water Air Soil Pollut.* **2001**, *128*, 85-96.
10. Bender, J.; Duff, M. C.; Phillips, P.; Hill, M. *Environ. Sci. & Tech.* **2000**, *34*, 3235-3241.
11. Duff, M. C.; Mason, C. F. V.; Hunter, D. B. *Can. J. Soil Sci.* **1998**, *78*, 675-683.
12. Kabay, N.; Demircioglu, M.; Yayli, S.; Guenay, E.; Yueksel, M.; Saglam, M.;
Streat, M. *Ind. Eng. Chem. Res.* **1998**, *37*, 1983-1990.
13. Zhao, P.; Romanovski, V.; Whisenhunt, D. W., Jr.; Hoffman, D. C.; Mohs, T. R.;
Xu, J.; Raymond, K. N. *Solvent Extr. Ion Exch.* **1999**, *17*, 1327-1353.
14. Madic, C.; Hobart, D. E.; Begun, G. M. *Inorg. Chem.* **1983**, *22*, 1494-1503.

15. Weigel, F. In *Handbook on the Physics and Chemistry of the Actinides*; Freeman, A. J., Keller, C., Eds.; North-Holland: Amsterdam, 1985; Vol. 3, pp 243-288.

16. Vizitiu, D.; Walkinshaw, C. S.; Gorin, B. I.; Thatcher, G. R. J. *J. Org. Chem.* **1997**, *62*, 8760-8766.

17. Ashton, P.; Koniger, R.; Stoddart, J. *J. Org. Chem* **1996**, *61*, 903-908.

18. Gans, P.; Sabatini, A.; Vacca, A. *Talanta* **1996**, *43*, 1739-1753.

19. Mustafina, A. R.; Skripacheva, V. V.; Kazakova, E. K.; Markarova, N. A.; Kataev, V. E.; Ermolaeva, L. V.; Habicher, W. D. *J. Incl. Phenom. Macrocycl. Chem.* **2002**, *42*, 77-81.

20. Morozova, Y. E.; Kuznetzova, L. S.; Mustafina, A. R.; Kazakova, E. K.; Morozov, V. I.; Ziganshina, A. Y.; Konovalov, A. I. *J. Incl. Phenom. Macrocycl. Chem.* **1999**, *35*, 397-407.

21. Batinic-Haberle, I.; Spasojevic, I.; Crumbliss, A. L. *Inorg. Chem.* **1996**, *35*, 2352-2359.

22. Bayada, A.; Lawrance, G. A.; Maeder, M.; Martinez, M.; Skelton, B. W.; White, A. H. *Inorg. Chim. Acta* **1994**, *227*, 71-77.

23. Kano, K.; Kitae, T.; Shimofuri, Y.; Tanaka, N.; Mineta, Y. *Chem. Eur. J.* **2000**, *6*, 2705-2713.

24. Connors, K. *Chem. Rev.* **1997**, *5*, 1325-1357.

25. Ashton, P.; Boyd, S.; Gattuso, G.; Hartwell, E.; Koniger, R.; Spencer, N.; Stoddart, J. *J. Org. Chem* **1995**, *60*, 3898-3903.

26. Rong, D.; D'Souza, V. *Tet. Lett.* **1990**, *31*, 4275-4278.

27. Submitted to *Chem. Comm.*

28. Cassol, A.; Di Bernardo, P.; Portanova, R.; Tolazzi, M.; Tomat, G.; Zanonato, P. *Inorganic Chemistry* **1990**, *29*, 1079-1084.

29. Cassol, A.; Di Bernardo, P.; Pilloni, G.; Tolazzi, M.; Zanonato, P. L. *Dalton Trans.* **1995**, 2689-2696.

30. Rudiger, V.; Schneider, H. *Chem. Eur. J.* **2000**, *6*, 3771-3776.

31. Alderighi, L.; Gans, P.; Ienco, A.; Peters, D.; Sabatini, A.; Vacca, A. *Coord. Chem. Rev.* **1999**, *184*, 311-318.

Environmental Implications: Metrology for Nanosized Materials

Editors
Clayton Teague
Barbara Karn

Chapter 14

Environmental Implications: Metrology for Nanosized Materials

Clayton Teague[1] and Barbara Karn[2]

[1]National Nanotechnology Coordinating Office, National Science
Foundation, 4201 Wilson Boulevard, Arlington, VA 22230
(cteague@nsf.gov)
[2]National Center for Environmental Research and Quality,
U.S. Environmental Protection Agency, 401 M Street, SW,
Washington, DC 20460 (Karn.Barbara@epa.gov)

Metrology supports industry by enabling the benefits of new products and processes to be measured and by stimulating new product development in the instrument sector in addition to raising productivity through improved process and quality control. Measurement also provides a foundation for a wide range of public goods, including environmental and human health protection. So, how do instrumentation and metrology and the environment specifically relate to one another? What is the importance of instrumentation and metrology to environmental and human health protection? What needs can instrumentation and metrology supply for nanotechnology related to the environment? How do the nanotechnology research goals for instrumentation and metrology and environmental protection coincide? The answers to these questions can be addressed under four topics that comprise metrology and the uses to which its results are applied—written standards, scientific instrumentation, validated measurement procedures, and measurement standards.

Protecting the environment relies on written standards. Limits are placed on emission of pollutants; air, to be healthy, must meet certain standards; both drinking water and surface waters have quality criteria; pollutants are measured and c ontrolled f rom l andfills. M etrology i s n ecessary t o e nable t he measurements that form the basis for deciding environmental standards. Currently, there are no standards for nano-sized m aterials i n t he e nvironment. Particularly in the air, nano-sized particulate matter may have an impact on human health. However, the inability to measure ambient nanoparticulates quickly, inexpensively, and accurately severely restrains research that would lead to intelligent standards for airborne nanoparticles that might effect human health. Waterborne nanoparticles may likewise need standards if they are found to be harmful to aquatic organisms.

Scientific instrumentation is essential in examining nanoscale materials and their interactions and impact on and in the environment. In addition to the need for instruments that detect and characterize nanomaterials, nanotechnology itself can form the basis for detection of other materials. New instruments using massively p arallel n anoscale sensor arrays could enable more sensitive, highly selective detection of environmentally important analytes, including both chemical compounds and biological organisms such as algae, bacteria, or viruses. Instrumentation is needed for monitoring nanomaterials in ambient environments in order to determine the fate, transport and transformations of these substances. Current metrology instrumentation for nanotechnology involves integrated use of biological principles, physical laws, and chemical properties. These three aspects are also integrated into environmental science, and any furtherance of the integrative science will also further fundamental research in the environment.

Validated measurement procedures are necessary for quality control in monitoring and analyzing environmental samples. If emissions standards are implemented, there must be reasonable means for effected industries and organizations to measure nanoparticles for compliance requirements. Measurement procedures also help the environment by increasing the efficiency of manufacturing processes, thereby producing less waste t hat c ould h arm t he environment.

Measurement standards are essential for research to proceed in nanotechnology related to the environment. If there is no standardization, there is no comparability among the numerous research laboratories involved in examining nano-scale science and its relation to the environment. Measurement standards are also needed for quality control during monitoring.

The importance of measurement in the environment at the nanoscale was discussed at a workshop on Nanotechnology Grand Challenge in the Environment. May 8-9, 2003. One of the five major discussion topics was Nanotechnology Applications for Measurement in the Environment as applied to sensors, monitors, models, separations, detection, fate and transport, data gathering and dissemination. The vision statement of this workshop group says: "The unique properties of nanoscale materials will enable the development of a new generation of environmental sensing systems. In addition, measurement science and technology will enable the development of a comprehensive understanding of the interaction and fate of natural and anthropogenic nanoscale and nanostructured materials in the environment."

Research needs were identified in five areas: (1) biological sensor technologies that are sufficiently stable to allow detection *in situ* on a continuous basis for high-density usage; (2) a general "array" for detection of a wide variety of potential analytes; (3) information concerning the diversity of chemical composition at the nanoparticles level, and t he t ransformations t hat o ccur a nd measurement techniques that distinguish the chemical composition of particle surface layers from the particle interior; (4) generic nanoscale assembly methods; and (5) advances in spectroscopic instrument technologies that allow rapid detection of low signal strength, while probing smaller volumes of a nanoparticulate sample.

Measurement is fundamental to the progress and quality of all scientific endeavors and engineering applications. It provides an underlying foundation for research in environmental and human health protection. Advances in metrology to both measure nanoscale materials and to use nanotechnology in measurement go hand in hand with advancing the protection of the environment and human health.

Chapter 15

Molecular-Dynamics Simulation of Forces between Colloidal Nanoparticles

Kristen A. Fichthorn[1,2] and Yong Qin[1,3]

[1]Department of Chemical Engineering, The Pennsylvania State University,
University Park, PA 16802
[2]fichthorn@psu.edu
[3]yqin@psu.edu

Introduction/Background

Nanoparticles hold great promise for a diverse array of materials applications, ranging from electronic circuits to bulk materials with novel mechanical properties to biological materials. Many applications involve colloidal nanoparticles, whose effective use in nanotechnology hinges on their selective assembly or their stabilization against aggregation. Various methods have been used to stabilize colloidal nanoparticles; however all involve dispersant molecules such as surfactants or polyelectrolytes. Not only do these dispersants alter the chemistry and physics of nanoparticle systems, but since they occupy a significant mass fraction of a nanoparticle system, they produce a tremendous waste stream during processing. An improved understanding of the forces between "bare" colloidal nanoparticles could lead to new and environmentally beneficial strategies for engineering colloidal nanoparticle suspensions.

Historically, the Derjaguin-Landau-Verwey-Overbeek (DLVO) theory has been used to describe electrostatic and van der Waals interactions in colloidal systems [1]. However, the assumptions of DLVO theory do not apply to nanoparticles. Further, recent studies suggest that forces that are not taken into account by DLVO theory, such as solvation and depletion, could be important in colloidal nanoparticle systems. From a theoretical point of view, it is now possible to simulate colloidal nanoparticles using large-scale, parallel molecular dynamics (MD). These studies can yield atomic-scale detail that is not currently accessible with experimental methods and they can be used to resolve the origins and magnitudes of forces between colloidal nanoparticles. The goal of this study is to apply MD simulations to investigate the relative magnitudes of solvation and van der Waals forces in a model colloidal nanoparticle system.

Materials/Methods

We use parallel MD to simulate two solid nanoparticles immersed in a liquid. The bulk solvent is simulated as 108,000 Lennard-Jones atoms in a cubic cell with periodic boundary conditions. To study the influence of particle size on interparticle forces, we simulate spherical nanoparticles of two different sizes. The small spherical nanoparticles are composed of 64 atoms and the average diameter is 1.67 nm. The large spherical nanoparticles are composed of 2048 atoms and the average diameter is 6.01 nm. The surfaces of the spherical nanoparticles are rough, as shown in Fig. 1. To investigate the influence of surface roughness, the nanoparticles are rotated so that they contact from a different angle and have different contacting surfaces. Finally, since the shape of a nanoparticle can influence solvation forces, we also study cubic nanoparticles, which have 2744 atoms arranged in a face-centered-cubic (fcc) structure, such that the contacting surfaces have the smooth and flat fcc(111) structure.

Particle-solvent interactions can further affect the solvation force between nanoparticles. In the simulations, nanoparticles can be either solvophilic or solvophobic. Solvophilic means that the solvent molecules have a stronger attraction to atoms in the nanoparticles than they do to themselves. To model this case, we set the solid-fluid attraction to be five times stronger than fluid-fluid attraction. For solvophobic nanoparticles, the interaction between the solvent molecules and the nanoparticle atoms is five times weaker than the solvent-solvent attraction.

Results/Discussion

Several existing theories can be used to calculate van der Waals forces. These theories rely on the assumption that the van der Waals attraction between two colloidal particles is the sum of pair-wise additive dispersion forces between the atoms comprising the objects. Using the assumption of a continuum solid, Hamaker [2] and Bradley [3] derived expressions for the van der Waals interaction between two macroscopic, spherical particles. In both of these formulae, the scaling of the interaction with particle separation is different than the $1/r^6$ scaling between two atoms. Since nanoparticles are an intermediate case between single atoms and macroscopic colloidal spheres, it is unclear whether these macroscopic theories can describe van der Waals interactions between them.

We calculated and compared van der Waals forces between the spherical nanoparticles using direct evaluation of the pair-wise dispersion forces, as well as with Hamaker's and Bradley's equations. Due to the small size and irregular surfaces of the nanoparticles, results with different theories vary considerably.

Both Hamaker's and Bradley's equations fail for short nanoparticle separations. The discrepancy between Hamaker's formula and direct evaluation arises from the irregular surface structure of the nanoparticles and Hamaker's equation predicts van der Waals forces well at separations greater than 3 fluid layers (about 1 nm). In both the small and large nanoparticle systems, Bradley's equation always underestimates the van der Waals force.

The free energy change of the nanoparticle system with particle separation is calculated from a variant of the thermodynamic integration (TI) method [4]. In the TI method, the derivative of the free energy with respect to a nanoparticle separation δ is calculated directly and then integrated to yield $\Delta A(\delta)$,

$$\Delta A_{ij} = \int_{\delta_i}^{\delta_j} d\delta \left\langle \frac{dU(\delta)}{d\delta} \right\rangle_\delta \qquad , \qquad (1)$$

where δ_i and δ_j are states of the system with different nanoparticle separations. Accounting only for solvation forces in the derivative of the potential U, we have

$$\Delta A_{ij} = \int_{\delta_i}^{\delta_j} d\delta \left\langle \hat{r}_{AB} \cdot \left(\hat{F}_{A,s}(\delta) - \hat{F}_{B,s}(\delta) \right) \right\rangle_\delta \qquad , \qquad (2)$$

where $\hat{F}_{A,s}$ and $\hat{F}_{B,s}$ are the forces acting on particles A and B due to the solvent and $\hat{r}_{AB}$ is the unit vector pointing from particle A to particle B. The ensemble-average quantity in the integral of Eq. (2) is the solvation force.

Solvation forces for solvophilic and solvophobic nanoparticles have been calculated for the different nanoparticle systems. In all the solvophilic nanoparticle systems, the solvation forces oscillate between attraction and repulsion. We find that the phase of the oscillations for the spherical nanoparticles is influenced by surface roughness and depends on their relative orientation. The oscillatory behavior is caused by the solvent's ordering near the surface. This effect is particularly evident for the cubic nanoparticles, which exhibit the strongest solvation forces. In Fig. 2, we show the solvation-force profile for the cubic nanoparticle along with the solvent-density profile from a slice of the simulation box surrounding the nanoparticles for a nanoparticle separation of 3.5σ. The dark red contours in the solvent-density plot indicate a higher solvent density than the bulk and solvent layering in the region between the two nanoparticles.

We find that solvation forces for solvophobic nanoparticles are always attractive. In this case, solvent molecules are repelled from the interparticle region and the density there is lower than the bulk density. Because the solvent-particle interaction is different from the solvophilic case, the solvophobic solvation forces are orders of magnitude smaller than the solvophilic solvation

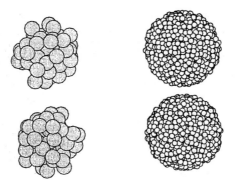

Figure 1: Snapshot of spherical 64- and 2048-atom nanoparticles.

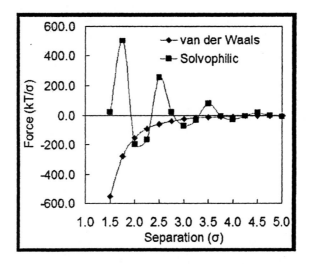

Figure 2: van der Waals and solvation forces for the solvophilic, cubic nanoparticle (left) and the solvent density profile for the cubic nanoparticle at a separation of about 3.5σ (right).

forces and van der Waals forces. Nevertheless, the attractive solvation forces enhance the attractive van der Waals forces between nanoparticles so they can aggregate more easily.

A comparison of solvation forces and van der Waals forces (assuming that the solid is Al_2O_3 and the solvent is Ar) for the cubic, solvophilic nanoparticles is shown in Fig. 2, where we see that the solvation force is comparable to the van der Waals forces. This indicates that solvation forces may be beneficial in preventing nanoparticles from aggregating and that stable nanoparticle dispersions may be achieved in suitable nanoparticle-solvent systems.

References

1. Israelachvili, J. N. (1992) "Intermolecular and Surface Forces", 2nd Ed., New York: Academic Press.
2. Hamaker, H. C. (1937) *Physica IV* **10**, 1058.
3. Bradley, R. S. (1932) *Phil. Mag.* **13**, 853.
4. Mezei, M. and Beveridge, D. L. (1986) *Ann. N. Y. Acad. Sci.* **482**, 1.

Chapter 16

Growth and Characterization of Single-Crystal Multilayer Nanostructures for Fast Ion Conduction

S. Azad, S. Thevuthasan, V. Shutthanandan, C. M. Wang, D. E. McCready, L. Saraf, O. A. Marina, and C. H. F. Peden

Pacific Northwest National Laboratory, 902 Battelle Boulevard, P.O. Box 999, Richland, WA 99352

Single-crystal multi-layered thin films of pure and mixed ceria and zirconia were heteroepitaxially grown using molecular beam epitaxy on yttria stabilized zirconia substrates. The growth of these films was monitored using *in-situ* reflection high-energy electron diffraction (RHEED). In addition, the *ex-situ* characterization techniques including X-ray diffraction (XRD), high-resolution transmission electron microscopy (HRTEM) and Rutherford backscattering spectrometry (RBS) along with ion channeling were used to further characterize these films.

Introduction

There has been considerable interest in solid oxide fuel cell (SOFC) devices since they provide relatively clean alternative energy to the conventional fossil fuels. The operation of all solid-state electrochemical devices including batteries, sensors and fuel cells, is essentially built on ion conduction in solid electrolytes. In order to improve the efficiency of electrochemical devices at lower operating temperatures, electrolyte materials with higher oxygen ionic conductivity at low temperatures need to be developed. Currently available electrolyte materials, *e.g.*, yttria-stabilized zirconia (YSZ), an almost exclusive oxygen ion-conducting electrolyte for modern solid oxide fuel cell devices, exhibit adequate oxygen-ion conductivity only at high temperatures (0.1 S/cm at 1275 K for YSZ). At lower temperatures, ionic conductivity in YSZ is significantly lower. Therefore, the operation of SOFC devices with YSZ electrolytes is limited to high temperatures only. At the same time, high operating temperatures significantly restrict the choice of materials for other components in the SOFC device. Several strategies are currently being explored in order to increase the ionic conductivity of the electrolyte and to reduce the operating temperature of the SOFC devices. The prevalent approach to increase the conductivity is based on the incorporation of different dopants, which preferentially increase the concentration of certain types of lattice defects in the electrolytes. In particular, cerium oxide doped with a divalent or trivalent oxide (in most cases, gadolinia or samaria) exhibits higher ionic conductivity, compared to YSZ, at temperatures around 1075 K [1-7].

It is well known that nanoscale materials often display properties that are very different compared to the base, coarse-grained, bulk materials. In particular, it has been recently demonstrated that a nanoscale lamellar structure of two different fluorides (calcium fluoride and barium fluoride) can exhibit significantly higher ionic conductivity along the interfacial directions at moderate temperatures [8]. In that arrangement, the ionic conductivity of barium fluoride was increased 100 to 1,000 times compared to that of a bulk crystal. If such a remarkable finding could be transferred into practice, it would provide the ability to design similar structures from oxygen ion conductors to enhance the performance of solid-oxide fuel cells, oxygen generators and membrane reactors at substantially reduced operating temperatures. In this study, we investigate a novel approach of increasing the ionic conductivity of doped ceria by artificially introducing nanoscale interfaces that are parallel to conduction.

As a first step, we have grown and characterized high quality single crystal multi-layered pure ceria and zirconia films on YSZ(111). The films were grown by oxygen plasma assisted molecular beam epitaxy (OPA-MBE) and were characterized by reflection high-energy electron diffraction (RHEED), x-ray diffraction (XRD), Rutherford backscattering spectrometry (RBS) and high resolution transmission electron microscopy (HRTEM). Since these high quality single crystal films, with minimum defects, have weak oxygen ionic conductivity, it is difficult to obtain useful information regarding the effects of nanoscaling on oxygen ion transport properties of these materials. In general, the oxygen ionic conductivity is significantly higher in doped ceria and zirconia films compared to pure films and the growth of these films with dopants is currently underway in our laboratory. In this short paper, the growth and characterization of the single crystal ceria/zirconia alternating multi-layer films is discussed. Details of the films with dopants, including ion conductivity measurements, will be described in a later publication.

Experimental

All growth and *in-situ* characterization measurements were carried out in a dual-chamber ultrahigh vacuum (UHV) system described in detail elsewhere [9]. The MBE chamber consists of three metal evaporation sources and an UHV compatible electron cyclotron resonance (ECR) oxygen plasma source, as well as RHEED for real-time characterization of epitaxial film growth. Ce and Zr rods (both 99.98% purity) were used as the source materials in separate electron beam evaporators. Growth rates of these films were monitored by means of quartz crystal oscillators (QCO) adjacent to the substrate. Polished (10×10×2) YSZ (yttria-stabilized zirconia) (111) single crystals were used as substrates for epitaxial growth of ceria-zirconia multi-layer films. YSZ substrates were ultrasonically cleaned in acetone and methanol prior to insertion into the dual-chamber UHV system through a load lock. Once in the MBE chamber, the substrates were cleaned by heating to 875 K for 10 min while being exposed to activated oxygen from the ECR plasma source at an oxygen partial pressure of ~1.5 x 10^{-5} Torr. The ceria-zirconia epitaxial films were grown by alternately evaporating Ce and Zr metal beams in the presence of a low-pressure oxygen plasma. The optimum experimental conditions that were used during the growth of single crystal zirconia-doped ceria in a previous study [10] were used to grow these films. In particular, the substrate temperature and the oxygen partial pressure were kept at 925 K and 2.0x10^{-5} Torr, respectively, during the growth. The rates of cerium and zirconium metal deposition were maintained between 0.02-0.03 nm/sec and the growth was monitored by *in-situ* RHEED.

The samples were further characterized by *ex-situ* XRD, RBS and HRTEM techniques. In the XRD analysis, the samples were measured on a high-resolution, double-crystal Philips X'Pert MRD system. The experimental diffraction data were analyzed using Jade (Materials Data, Inc., Livermore, CA) and compared with the Powder Diffraction File PDF-2 database (International Centre for Diffraction Data, Newtown Square, PA). RBS work was performed in the Environmental Molecular Science Laboratory (EMSL) accelerator facility, which is described elsewhere [11]. RBS and channeling measurements were carried out at a sample temperature close to 300K using 2.04 MeV He^{+} ions at normal incidence. The SIMNRA simulation program [12] was used to model the experimental RBS spectra, and determine the thickness and stoichiometry of the films. Finally cross section thin foils were prepared by standard tripod wedge polishing, followed by Ar-ion beam thinning for high HRTEM measurements and these measurements were carried out on a Jeol JEM 2010 microscope with a routine point-to-point resolution of 0.194 nm.

Results And Discussion

A high quality 7 nm single crystal ceria layer was first grown on a clean YSZ(111) substrate. Fig. 1 shows typical RHEED patterns obtained in the [-110] (left panels) and [-211] (right panels) azimuthal directions from the YSZ substrate (top panels) and a pure ceria film grown on YSZ (bottom panels). The substrate patterns are of good quality, showing streaks rather than an array of spots, indicating reasonably flat surfaces with large terrace widths. The RHEED patterns obtained from the epitaxial ceria film also indicate reasonably well-ordered surfaces showing RHEED patterns with low background, homogenous intensity along the streaks, and clear visibility of Laue zones.

The RHEED patterns for the films persisted from the beginning of the growth, suggesting that these films grow in a reasonably laminar fashion on YSZ. The streak spacings of the films are qualitatively similar to that of the bulk spacing of CeO_2 suggesting that the films have achieved the bulk spacings of CeO_2 as expected. In addition, there is no evidence for surface reconstruction under these growth conditions. The RHEED patterns reveal that epitaxial films have the same orientation relationship as YSZ. However, during the growth of 7 nm pure zirconia layer on top of the ceria film, half order streaks were observed in addition to the primary streaks observed during ceria growth. This may indicate the presence of a surface reconstruction for zirconia. The half order streaks persisted until the end of 7 nm zirconia growth and disappeared during the growth of next ceria layer. Since a cubic structure for pure zirconia is not stable, it is possible that this leads to a surface reconstruction, although we also note the stress in the films associated with a lattice mismatch to maintain the cubic fluorite structure that was obtained in XRD measurements discussed below. Appearance and disappearance of half order streaks during zirconia and ceria growth, respectively, continued throughout the growth of the film that contained 10 alternating layers of ceria and zirconia.

The XRD spectrum obtained for the hetero layers of CeO_2 and ZrO_2 on the YSZ(111) substrate is shown in Fig. 2. The observed positions of the $CeO_2(111)$ and $CeO_2(222)$ peaks at $2\theta \sim 28.5$ and 58 closely match with the reference data and the absence of any other reflections in the pattern indicates that the CeO_2 film was single phase. Similarly, the positions for the $ZrO_2(111)$ and $ZrO_2(222)$ also matched the reference data from the cubic fluorite zirconia structure, and these peaks were almost superimposed on the much sharper substrate YSZ peaks. XRD results reveal that the thin film, composed of hetero-layers of CeO_2 and ZrO_2, has a high crystalline quality, and also that the film has single phase CeO_2 and ZrO_2.

In Fig. 3 (a), we present the channeling and random spectra of epitaxially grown hetero-layers of CeO_2 and ZrO_2 thin films on YSZ(111). The arrows in the figure indicate the energetic positions expected for backscattering from Ce,

Substrate - YSZ

[-110] **[-211]**

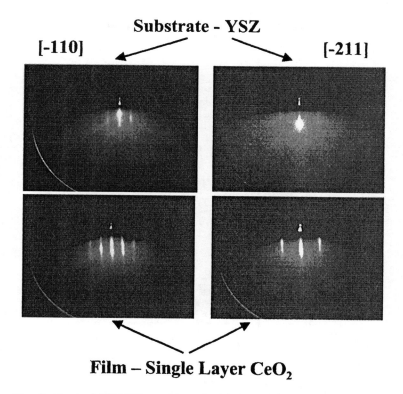

Film – Single Layer CeO₂

Fig. 1: Typical RHEED patterns obtained in the [-110] (left panels) and [-211] (right panels) azimuthal directions from the YSZ substrate (top panels) and a pure ceria film grown on YSZ (bottom panels).

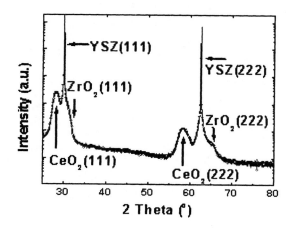

Fig. 2: XRD data for hetero-layers of CeO₂ and ZrO₂ thin films on YSZ(111).

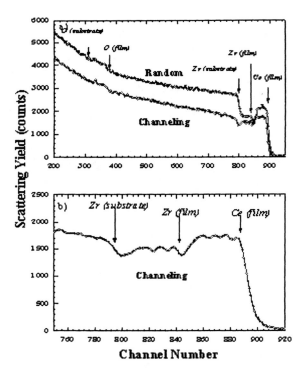

Fig. 3: Channeling and random RBS spectra for epitaxially grown hetero layers of CeO$_2$ and ZrO$_2$ thin films on YSZ(111). Incident energy of the He$^+$ beam was 2.04 MeV and the scattering angle was 150°.

Zr and O atoms. It is clear from the figure that there is some channeling in the film; hence, it has some degree of crystallinity. The multi-layered structure of the films is also very clear in Fig. 3(b), an enlargement of a region of the channeling spectrum shown in Fig. 3(a). It is well established that the pure CeO_2 and Zr-doped CeO_2 films on YSZ substrates show misfit-dislocations at the interface [13]. Due to these misfit-dislocations, the channeling spectrum from these films shows an extra peak at the interface [14]. As observed in the TEM results discussed below, these epitaxially grown hetero-layers of CeO_2 and ZrO_2 thin films on YSZ(111) studied here show misfit dislocations at the interfaces as well. The combination of this disorder and the relatively poor depth resolution ($\sim$ 20 nm) of RBS, results in minimum yields of Ce and Zr that are higher than expected for thicker films of good crystalline quality.

Fig. 4(a) is a low magnification TEM image of a film composed of 5 layers of ZrO_2 and 5 layers of CeO_2 grown in an alternative sequence. The thickness of each layer appears to be uniform, as evidenced by the straight interface between different layers. The inset of Fig. 4(a) shows the selected area electron diffraction (SAED) of the film, demonstrating that both ZrO_2 and CeO_2 layers are single crystals, and that those layers maintain an epitaxial orientation relationship of $[110]ZrO_2//[110]CeO_2$ and $(002)ZrO_2//(002)CeO_2$. One interesting aspect of the SAED is the existence of the otherwise forbidden diffraction spots for the fluorite structure, such as spot (001) as (even, even, odd). The origin of these forbidden spots will be discussed in a separate paper. The lattice mismatch between ZrO_2 and CeO_2 layers is also clearly revealed by the separation of high order diffraction spots. The lattice mismatch between the ZrO_2 and CeO_2 layers is mainly relieved by interface misfit dislocations, as shown in the HRTEM image of Fig. 4(b). In addition, a few dislocations are also visible in the interior of the film.

Conclusions

High quality single crystal multi-layered pure ceria and zirconia films were grown using molecular beam epitaxy on YSZ(111). These films were characterized using *in-situ* reflection high-energy electron diffraction (RHEED) and *ex-situ* X-ray diffraction (XRD), high-resolution transmission electron microscopy (HRTEM) and Rutherford backscattering spectrometry (RBS) along with ion channeling. The RHEED patterns obtained from the epitaxial multi-layered films indicate reasonably well-ordered surfaces. XRD, channeling and HRTEM measurements, also indicate that these films are single phased and high quality single crystals.

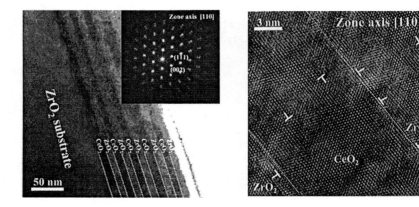

Figure 4. Low magnification TEM image showing the general structure of the multi-layer film (a). HRTEM image showing the interface structures between different layers (b).

Acknowledgements

This research was supported in part by the Division of Chemical Sciences, Office of Basic Energy Sciences, U.S. Department of Energy. The experiments were performed in the Environmental Molecular Sciences Laboratory, a national scientific user facility located at Pacific Northwest National Laboratory (PNNL), and supported by the U.S. Department of Energy's Office of Biological and Environmental Research. PNNL is a multi-program national laboratory operated for the U.S. DOE by Battelle Memorial Institute under contract No. DE-AC06-76RLO 1830.

References

1. Z. Tianshu, P. Hing, H. Huang, and J. Kilner, Solid State Ionics **148** (2002) 567.
2. G.Y. Meng, Q.X. Fu, S.W. Zha, C.R. Xia, X.Q. Liu, and D.K. Peng, Solid State Ionics **148** (2002) 533.
3. M. Boaro, A. Trovarelli, J.H. Hwang, and T.O. Mason, Solid State Ionics **147** (2002) 85.
4. H. Zhao and S. Feng, Chem. Mater. **11** (1999) 958.
5. E. Ruiz-Trejo, J.D. Sirman, Y.M. Baikov, and J.A. Kilner, Solid State Ionics **113-115** (1998) 565.
6. W. Huang, P. Shuk, and M. Greenblatt, Solid State Ionics **113-115** (1998) 305.
7. T. Ohashi, S. Yamazaki, T. Tokunaga, Y. Arita, T. Matsui, T. Harami, and K. Kobayashi, Solid State Ionics **113-115** (1998) 559.
8. N. Sata, K. Eberman, K. Eberl, and J. Maier, Nature **408** (2000) 946.
9. S.A. Chambers, T.T. Tran, and T.A. Hileman, J. Mat. Res. **9** (1994) 2944.
10. Y. Gao, G.S. Herman, S. Thevuthasan, C.H.F. Peden, and S.A. Chambers, J. Vac. Sci. Technol. A **17** (1999) 961.
11. S. Thevuthasan, C.H.F. Peden, M.H. Engelhard, D.R. Baer, G.S. Herman, W. Jiang, Y. Liang, and W.J. Weber, Nucl. Instr. Meth. A **420** (1999) 81.
12. SIMNRA User's Guide. Edited by M. Mayer, Max-Plank-Institut fur Plasmaphysik, Germany (1997).
13. C.M. Wang, S. Thevuthasan, and C.H.F. Peden, J. Am. Ceram. Soc. **86** (2003) 363.
14. V. Shutthanandan, S. Thevuthasan, Y.J. Kim, and C.H.F. Peden, Mat.Res.Soc.Symp.Proc. Vol. **640** (2001).

Chapter 17

Single-Particle Mass Spectrometry of Metal-Bearing Diesel Nanoparticles

Donggeun Lee, Art Miller, David Kittelson, and Michael R. Zachariah*

Department of Mechanical Engineering, University of Minnesota, 111 Church Street Southeast, Minneapolis, MN 55455
*Corresponding author: mrz@me.umn.edu

Introduction/Background

It is known that diesel engines produce a tri-modal size distribution of DPM (diesel particulate matter), (1). The number of nano-size particles i.e. those in the "nuclei" mode, is quite variable and can be attributed in many cases to the self-nucleation of volatile species during the dilution and simultaneous cooling of the exhaust. In some cases nanoparticles may also form prior to dilution (2, 3).

The dynamics of particle formation during and after combustion is a topic of much interest and investigation. A recent summary of the diesel combustion process (4) describes how the fuel jet quickly disintegrates as it exits the nozzle and vaporizes as it entrains hot air, subsequently forming a teardrop-shaped cloud of fuel-vapor and air mixture with a diffusion flame at its periphery. Particles of soot originate from pyrolysis of fuel in the fuel-rich region of this diffusion flame. These primary soot particles are spherical and consist of many layers of carbon atoms and/or multi-atom platelets. As the piston moves downward, rapid adiabatic cooling causes these primary particles to agglomerate. At this stage, if there are any species (such as metals) with very high concentrations, they may self-nucleate if the rapid cooling drives their saturation ratios high enough. As the resulting aerosol travels through the exhaust system, further cooling causes condensation of other species onto the particles. When the aerosol exits the tailpipe it again cools rapidly and other more volatile species condense suddenly and/or self-nucleate into nanoparticles. The resulting mixture of gases and particles is what we call DPM.

Both acute and chronic health effects have been associated with DPM. One theory holds that the potentially large number of ultrafine particles and their characteristic high lung-penetration efficiency may play a role in this correlation (5), and it is also possible that the presence of metals may be a contributor. To address these issues, the UMN and NIOSH are conducting a joint investigation to characterize the metal content of diesel nanoparticles and this report summarizes preliminary findings. The research aims at providing insight into the formation of metal-rich nanoparticles via self-nucleation as well as the overall distribution of metal across a wider range of particle sizes due to condensation and coagulation.

Methods/Materials

The source of DPM for this work was a three cylinder, 1.5 liter Isuzu engine powering a genset. Engine speed was constant at 1800 rpm and the engine was run at two conditions i.e. no-load and 6kW load, with fuel flow rates of 1.0kg/hr and 2.25kg/hr respectively. The fuel used for all tests was #2 diesel fuel with nominal sulfur level of 350 ppm. The fuel was doped with iron using small amounts of ferrocene i.e. $(C_5H_5)_2Fe$, blended with the fuel to achieve the desired doping level. DPM samples were drawn from the tailpipe by a stainless steel probe inserted through the pipe wall and directed into the exhaust stream. The sample was subsequently diluted approximately 10 to 1 with dry air.

The single-particle mass spectrometer (SPMS) used in this study is shown in figure 1 and described in detail in reference 6. The primary components of the system are an aerodynamic lens inlet (7), a two-stage differential pumping system, a free-firing dissociation/ionization laser and optics system, a linear time-of-flight mass spectrometer, and a data acquisition system composed of a high-speed digital storage oscilloscope and PC.

The SPMS generates particle mass spectra akin to that in figure 2. Note that although there are some molecular fragments (in this case Cx, N2, O2), the particles are generally ablated to their elemental constituents. Thousands of such spectra were collected during the course of this work. Each spectrum is unique, as is each particle. Along the lines of earlier work (8), customized software was developed to help sort and analyze the numerous individual particle spectra.

Since the SPMS does not have a means for measuring the size of particles that are ablated, a method was developed to correlate the size distribution of the inlet aerosol to the range of signal intensities of the many particle mass spectra. The inlet aerosol size-distribution was first measured with the SMPS, and corrected using an empirically derived transmission efficiency for the aerodynamic lens. Next, the sizes of all sampled particles were estimated. This

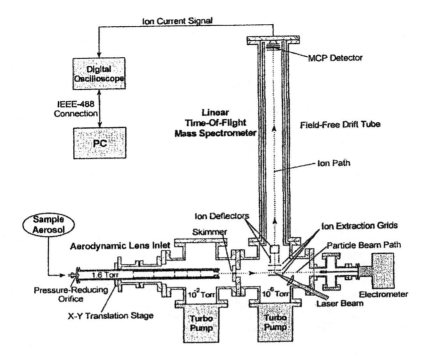

Figure 1. Schematic of SPMS

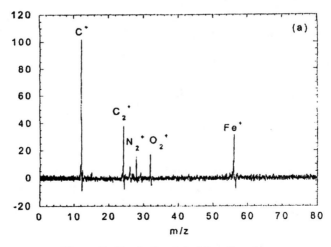

Figure 2. Single Particle Mass Spectrum

was done by conducting a series of tests to correlate the SPMS signal intensity to the size of known particles (see also reference 9). From these tests we derived a non-linear correlation factor for estimating particle size from SPMS signal intensity. We used that factor to calculate a volume-equivalent particle diameter for each sampled particle. The matching of the two resulting size distributions (figure 3) is indicative of the accuracy of using this approach for diesel particles.

Results/Discussion

When the fuel is doped with 20ppm iron, the SMPS-measured particle size distribution contains a distinct nuclei mode which increases with engine load/fuel flow rate (figure 4).

Due to the great variety of particles in the aerosol, the resulting single particle mass spectra vary considerably. To help sort the spectra we counted on previous knowledge of particle morphology and divided the particles into three classes as follows. Note that the values of C, H, and Fe used were the areas under the elemental peaks of the individual mass spectra.

"Organic Carbon" particles are defined as those where H/C >1. They are typically nanoparticles containing fuel or oil residues rich in organic carbon i.e. rich in hydrogen.

"Elemental Carbon" particles are those where H/C < 1. These are particles containing greater relative amounts of elemental carbon (typically agglomerates of primary carbon spherules).

"Pure iron" particles are those where Fe/C>15.   These are the nanoparticles of iron (self nucleated in the engine).

To verify this for our case we took DPM samples for complimentary analyses i.e. transmission electron microscopy and energy dispersive spectroscopy (TEM/EDS), using a low-pressure cascade impactor (LPI), (10). Figure 5 shows a TEM image which contains all three of the above mentioned particle types.

Figure 6 shows the size distribution for the above three particle types based on numerous mass spectra, for the 6kW engine load. Organic carbon particles are more prevalent in the smaller size ranges, which is consistent with a previous observation (11) indicating that the amount of organic carbon on DPM decreases as particle size increases. Note that about 8% of the particles detected are pure iron nanoparticles and that the elemental carbon particles are distributed similarly to the accumulation mode shown in figure 4. The latter suggests that the coagulation of primary carbon particles into agglomerates does not change significantly with metal doping.   However, the absolute number of these accumulation mode particles was shown to decrease as a result of doping. This supports earlier work (2) and suggests that the presence of metals during the

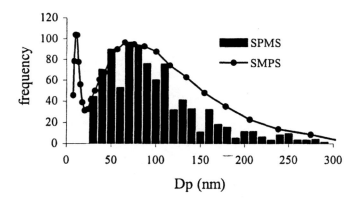

Figure 3. Size Correlation of SMPS and SMPS data

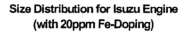

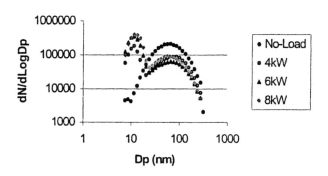

Figure 4. Size Distribution with 20ppm Doping

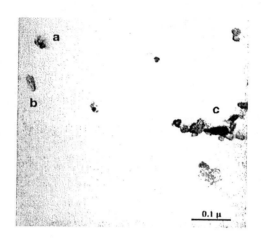

Figure 5. a) Fe, b) HC, and c) Agglomerate particles

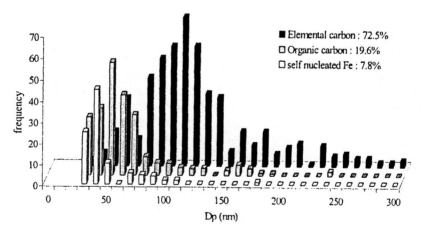

Figure 6. Size and Frequency of Three Particle Types

combustion process may enhance the oxidation of carbon. This is reflected in figure 4 where the higher presence of iron in the system i.e. at higher loads and thus higher fuel flow rate, is shown to reduce the numbers of accumulation-mode particles. That figure also shows that reducing the load and fuel flow causes the iron nanoparticles to virtually disappear as expected, and the elemental carbon particles to increase in number.

Figures 7a-b illustrate the analysis of numerous spectra to observe trends in the amount of various elements in each particle. Each dot on a graph represents the area under one elemental peak from one unique mass spectrum (i.e. one particle). The x-axis reflects the size of the particles analyzed, i.e. the volume-equivalent diameter calculated using the correlation factor mentioned above. The values on the y-axis reflect the amount of each element determined by calculating the area under that elemental peak. For carbon, we sometimes had more than one peak (due to molecular fragments) and in that case the area under all carbon-containing peaks would be summed to give the amount of carbon for that particle.

The SPMS spectra summarized in figures 7a-b show the fraction of iron in the particles at low and high engine load when using fuel doped with 20ppm iron. The figures clearly show the increase in absolute iron levels at high load (which reflects the doubling of fuel flow rate). At low load, iron fraction increases non-linearly for smaller particles, which suggests that iron condensation onto carbon particles plays a key role in particle formation, since that would favor smaller particles due to their higher surface to volume ratio. Contrast this with the data of figure 7b for higher engine load, which still shows the non-linear behavior but with significant data scatter. This increased scatter is believed to be a result of a trend toward coagulation dominance in the system, i.e. an increasing role played by nucleation and subsequent coagulation of pure iron nanoparticles. At high load, the quantity of Fe more than doubles due to increased fuel flow rate. This increases the saturation ratio of iron vapor in the combustion chamber and the nuclei mode particles subsequently grow larger and more numerous (as in figure 3). Figure 7b shows that for the higher engine load there is a group of data with Fe/(C+Fe) ratios of unity, representing those pure iron nanoparticles. Figure 7a does not show such particles, since at that condition they are too few and too small to be detectable.

Conclusions:

- Using 20ppm iron doping, self-nucleated iron particles are detectable by both SMPS and SPMS.
- As iron content increases, the iron nanoparticles grow in size and number.

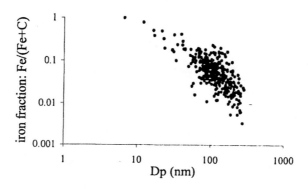

Figure 7a. Iron Fraction at Low Engine Load

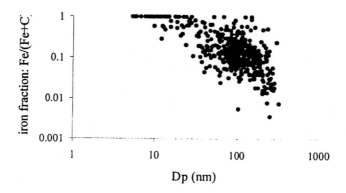

Figure 7b. Iron Fraction at High Load

- At low engine load, (or lesser iron availability), the proportion of iron increases non-linearly for small particles, suggesting condensation of iron onto particles.
- When more iron is available, self-nucleated particles increase in number and the amount of iron on larger particles is more variable, suggesting coagulation.

References:

1. Kittelson, D., 1998. "Engines and Nanoparticles: A Review". J. Aerosol Sci. Vol. 29, No. 5/6, pp. 575-588.
2. Du, C. J., Kracklauer, J., Kittelson, D., 1998. "Influence of an Iron Fuel Additive on Diesel Combustion". SAE Tech Paper 980536 pp. 1-13.
3. Skillas, G., Qian, Z., Baltensperger, U., Matter, U., Burtscher, H., 2000. "Infleunce of Additives on the Size Distribution and Composition of Particles Produced by Diesel Engines". Combustion Science and Technology, v 154, n 1, pp259-273.
4. Dec, J. E., 1997. "A Conceptual Model of DI Diesel Combustion Based on Laser-Sheet Imaging", SAE paper 970873.
5. Utell, M.J., Frampton, M.W., "Acute Health Effects of Ambient Air Pollution: The Ultra-fine Particle Hypothesis". Journal of Aerosol Medicine, 13, pp355-359, 2000.
6. Mahadevan, R., Lee, D., Sakurai, H., Zachariah, M., 2002. "Measurement of Condensed Phase Reaction Kinetics in the Aerosol Phase Using Single Particle Mass Spectrometry". Journal of Physical Chemistry A. 106, 11083.
7. Liu, P., Zeimann, P. J., Kittelson, D. B., and McMurry, P. H., 1995. "Generating Particle Beams of Controlled Dimensions and Divergence: II. Experimental Evaluation of Particle Motion in Aerodynamic Lenses and Nozzle Expansions." *Aerosol Science and Technology*, vol. 22, pp. 314-324.
8. D.J. Phares, K. P. Rhoads, A.S. Wexler, D.B. Kane, M.V. Johnston, "Application of the ART-2a Algorithm to Laser Ablation Aerosol Mass Spectrometry of Particle Standards", Analytical Chemistry (2001) 73, 2338-2344.
9. D.B. Kane, B. Oktem, and M.V. Johnston, "Nanoparticle Detection by Aerosol Mass Spectrometry", Aerosol Science and Technology (2001) 34, 520-527.

10. Hering, S., Friedlander, S., Collins, J. and Richards, W., 1979. "Design and Evaluation of a New Low-Pressure Impactor (part 2)". Enviro Sci and Tech Vol 13, Number 2, pp 184-188.

11. Sakurai, H., Tobias, H. J., Park, K., Zarling, D., Docherty, K., Kittelson, D. B., McMurry, P.H., Ziemann, P.J., "On-line measurements of diesel nanoparticle composition and volatility". Atmospheric Environment. v 37 n 9-10 March 2003. p 1199-1210.

Environmental Applications: Sensors and Sensor Systems

Editors
Susan L. Rose-Pehrsson
Pehr E. Pehrsson

Chapter 18

Environmental Applications: Sensors and Sensor Systems: Overview

Nanotech-Enabled Sensors and Sensor Systems

Susan L. Rose-Pehrsson and Pehr E. Pehrsson

Chemistry Division, Naval Research Laboratory, 4555 Overlook Avenue, SW, Washington, DC 20375–5342

The introduction and incorporation of nanomaterials and nanotechnology offers the potential for revolutionary transformations in the capabilities of sensors and sensor systems. Nanoscale sensor technologies rely on the unique physical and chemical properties conferred by their size. Intelligent use of such materials will permit the design of light, compact, low power, and exquisitely sensitive devices. Even old detection methods are being enhanced with the use of nanoscale materials. New sensor designs will exploit the inherent, often unique chemical and physical properties of nanoscale materials, such as engineered band structures, electrical conduction, mechanical resonance, quantum effects, optical properties, large surface to volume ratio, and small mass. For example, the large surface to volume ratio of a nanoparticles translates into large changes in the electrical conductivity or mechanical resonance of the nanostructure when very small quantities of analyte adsorb on its surface. The ability to manipulate the size of the particles, and especially their surface chemistry, therefore makes them amenable to designed modifications and optimization for particular analytical tasks. These qualities also make sensors based on nanotechnology especially well suited for arrays with different chemistries and transduction mechanisms, thus enhancing their selectivity and ability to discriminate analytes in complex chemical environments. These characteristics suggest that sensors will be increasingly flexible and customizable to specific tasks as research provides suites of suitable chemistries and designs built upon common architectures. Such a sensor can be based on varying numbers of these structures, from a single particle, such as a carbon nanotube, to an amalgamation of many particles, e.g. metal beads coated with an organic layer to form a 2-dimensional layer. Other sensors rely less on the specific size-dependent properties, but still benefit from using nanoscale particles optimized for other desirable device properties, e.g. low power consumption.

Traditional sampling and detection methods for environmentally important species can be time consuming, costly and require hazardous materials which themselves contribute to the environmental waste problem and create a logistical burden of operation. Nanotechnology may reduce or even eliminate the need for such reagents and can in some cases even enable the use of ambient species as part of the analytical process.

Unique sensor materials are being developed that result in small and inexpensive sensors with novel sampling mechanisms. Both optical and electronic-based nanocrystal sensors are being investigated because these novel materials offer unusual and useful properties. Likewise, carbon nanotubes possess an extraordinary range of electronic, thermal, and structural properties. These properties depend on the detailed physical and chemical structure of the tube, so small changes can dramatically affect them and offer a number of possible detection pathways.

The lines between chemical and biological nanosensors are often blurred, with chemical sensors sometimes used for measuring biological phenomena and vice versa. An excellent example of this is shown in Greenbaum's research where naturally occurring microorganisms are used as biosensors for chemical detection in drinking water. Researchers are increasingly combining the two fields in powerful new ways, especially as they look to nature for clues and ideas in developing novel nanostructures and sensors.

Nanotechnology is enabling sensitive new devices. Micromachined cantilevers-based sensors have a significant advantage in the absolute sensitivity achievable. Novel coating methods are making these sensors more robust and reproducible. This field is expected to advance quickly and produce some unique methods for detecting many chemical and biological species of interest.

In addition, old methods are being used and combined in new ways to miniaturize all aspects of sensor systems. One example is lab-on-a-chip technology, which permits transport of laboratory analytical methods to the field. The small size of the detection systems reduces sampling size, detection time, reagent quantities and laboratory waste products. This detection platform will need novel detectors with enhanced sensitivity, and nanotech-enabled devices are expected to provide many such devices.

Many analytes of environmental importance are now being investigated with these and other nanotech-enabled sensors. These unique sensors and sensor systems will permit simultaneous monitoring and remediation via real-time, on-line sampling, thus allowing timely prevention of environmentally hazardous conditions before they become serious problems.

Eight papers were invited to contribute to the Nanotech-Enabled Sensors and Sensor Systems Session of the Nanotechnology and the Environment Symposium. The papers range from basic research to development of specific chemical and biological sensors. In the following seven chapters, experts in the field share some the highlights of their work in this area. The papers are varied and describe both chemical and biological sensors. Nanotech-enabled sensors and sensor systems are in their infancy. The next decade will certainly see an explosion in their capabilities, hopefully accompanied by the ability to economically and reliably create these structures and effectively use the materials for the benefit of the environment.

Chapter 19

Sorptive Properties of Monolayer-Protected Gold Nanoparticle Films for Chemical Vapor Sensors and Arrays

Jay W. Grate*, David A. Nelson, and Rhonda Skaggs

Pacific Northwest National Laboratory, 902 Battelle Boulevard, P.O. Box 999, Richland, WA 99352
*Corresponding author: jwgrate@pnl.gov

Introduction

Nanoparticles and nanoparticle-based materials are attracting great interest for their unique properties and potential for application in diverse areas. Monolayer-protected nanoparticles (MPNs) are of particular interest because they can be taken up in solution, synthetically modified, or cast into thin films.[1-4]

On chemical microsensors, these materials act as sorptive films. Films of gold MPNs consist of gold particles separated from one another by the protective monolayers on their surfaces. These monolayers represent insulating layers of molecular dimensions and potential vapor sorption sites. Snow and Wohltjen described films of gold MPNs as Metal-Insulator-Metal-Ensembles (MIME) in recognition of the nanostructure of the films.[5] Initial investigations by a number of groups have confirmed the potential of MPN-based films for chemical vapor sensing.[5-12]

The response of a chemical microsensor with a sorptive coating is dependent on the amount of vapor sorbed by the film and the transduction of that sorbed vapor into a measurable change in the sensors output signal. Detection limits then depend on the signal relative to the noise. As a result, the detectability of vapors with such sensors depends on the signal-to-noise per sorbed vapor molecule. Accordingly, the sorptive properties of films of MPNs are of interest in understanding and developing nanoparticle-based sensors and sensor arrays for chemical vapor detection. Are these materials containing large mass percentages of gold and much smaller mass percentages of organic thiol effective at sorbing organic vapors, and how do they compare with sorptive polymers?

In this study, the thickness shear mode (TSM) device has been used to investigate the vapor uptake properties of several readily prepared gold MPN materials.*(12)* It is demonstrated that many, but not all, MPN-based sensing layers provide rapid and reversible uptake of vapors, as is desirable for vapor sensors. Sorptive selectivity varies with the monolayer structure, which provides a basis for sensing material design and use in sensor arrays. The sorptive properties are compared with more common polymeric sorptive materials so that conclusions can be reached regarding the relative roles of sorption and transduction in MPN-coated sensor performance.

Materials and Methods

A number of nanoparticle materials were synthesized using the two phase approach first described by Brust*(2)* and used subsequently by Wohltjen and Snow*(5)*. Nanoparticle materials were also synthesized by the single phase methanol/water synthesis developed previously by Brust*(1)* and used by others.*(7)* The nanoparticle materials were not fractionated and were used as isolated.

Nanoparticle materials were applied to one surface of a 10 MHz TSM device by spray coating solutions of the nanoparticles in either dichloromethane or methanol, depending on their solubility. Vapor sensors were always cleaned in a UVO Cleaner (Model 342, Jelight Company, Inc, Irvine CA) unit prior to film application. Vapor sorption measurements were made on TSM devices coated with an amount yielding approximately 10 kHz frequency shifts (an indication of film "thickness").

10 MHz TSM devices were operated with an oscillator card and the frequency signal was measured using a Hewlett-Packard 53131A High Performance Universal Counter with a medium stability time base, with data transferred to a microcomputer by GPIB. The frequency counter was controlled and data were logged using LabVIEW® software (National Instruments, Austin, TX). In addition, a Hewlett-Packard 4194A Impedance Analyzer was used to measure the motional resistances before and after coating crystals, and to determine the effect of coating amount on motional resistance. It was confirmed by these measurements that the coated TSM devices would function as gravimetric sensors.

Coated TSM devices were mounted in a flow cell with gas inlet and outlet ports, maintained at 298K, and exposed to calibrated vapor streams. The test vapors were generated from bubbler sources with dry nitrogen carrier gas, and further diluted with additional carrier gas using an automated vapor generation system. The sensors were exposed to 10 minute intervals of each vapor concentration followed by 15 minutes of carrier gas for recovery before the next concentration of the same vapor. Data points were typically collected every 2 seconds but longer intervals were sometimes used.

Results and Discussion

Nanoparticle materials. Dodecanethiol- and several arenethiol-protected gold nanoparticle materials were prepared for investigation; the thiols included dodecanethiol, chlorobenzenethiol, trifluorobenzenethiol, benzenethiol, and hydroxybenzenethiol. Nanoparticle formation was confirmed by high resolution transmission electron microscopy (TEM). Typical nanoparticle core sizes were in the range of 1.6 to 6 nm with about 3 nm most common.

Vapor sorption behavior. The test sensors were evaluated against nine concentrations each of four vapors: n-hexane, toluene, 2-butanone, and 1-butanol. The vapors were selected to represent a diverse set with nonpolar hexane, more polarizable toluene, basic and dipolar 2-butanone and hydrogen-bond acidic 1-butanol. These test vapors provide a preliminary probe of differences in behavior and chemical selectivity among the nanoparticle materials. Sensor responses to the test vapors were generally rapid and reversible, with exposures leading to steady state or near steady state responses within the six-minute exposure period. However, the responses of hydroxybenzenethiol-coated nanoparticle films were much slower and will not be further considered here.

Calibration curves for the four vapors on the selected coatings were determined. For gravimetric sensors at a fixed temperature, these represent sorption isotherms. These calibration curves range from mostly linear to varying degrees of downward concavity. The calibration curves for the benzenethiol-protected nanoparticle TSM sensor were the most curved in our data set.

Comparisons with sorptive polymers. Polymers are widely used in chemical sensing and their sorptive properties have been examined in detail. We have previous data for the sorption of hexane, toluene, and 2-butanone by poly(isobutylene), PIB, and poly(epichlorohydrin), PECH, on 10 MHz TSM devices.*(13)* These polymers have been used in a variety of vapor sensing studies, and represent simple prototypical sorptive polymers whose structures and sorptive properties have been described in detail elsewhere.*(13-16)*

Poly(isobutylene) an alkane based low-polarity polymer interacting by dispersion forces. In this regard it can be considered similar in interactive properties to the alkane chains of dodecanethiol. Poly(epichlorohydrin) is more dipolar due to the chloro-substitution. The arenethiol-based nanoparticle materials are expected to be polarizable and if halogenated, dipolar.

The sorptive data for the polymers were compared with the corresponding data for the nanoparticle films. This analysis compared the sorptive properties on a mass-of-vapor-sorbed per mass-of-sorbent-material. It was evident from the results that the best nanoparticle films for a given vapor were less sorptive on a per-mass basis than the best sorptive polymer considered, the polymer being better by a factor of 2 to 2.5.

However, it should be noted that films of equal mass but with different material densities will have different thicknesses. Vapor sorption is typically quantified on the basis of the partition coefficient, where the concentration of vapor in the sorbent phase is in g/L. Thus, the partition coefficient describes sorption on a mass-*per-volume* basis.

The densities of the polydisperse MPN materials as thin films are not known; however estimates were made using the mass compositions from thermogravimetric analysis and the bulk densities of gold metal and the thiol starting material. Although MPN materials are typically 75-90% metal by mass and thus only 10 to 25% organic by mass, they are 70-90% organic material by volume and only 10-30% metal by volume. Thus, despite the low mass percentage of sorptive organic insulating material, these materials are actually primarily sorptive organic material by volume.

Using the estimated film densities, we calculated preliminary estimates for partition coefficients. From these estimates, it was found that on a per-volume basis, the sorption of organic vapors by these MPN materials is of the same order of magnitude as the polymers at the test concentration.

Discussion.

Vapor sorption by the sensing film is a fundamental influence on sensor response. Regardless of the tranduction mechanism ultimately used for sensing, information on vapor uptake is useful in predicting and interpreting sensor response behavior. Slow uptake will lead to slow response times regardless of the transduction mechanism. Gravimetric calibration curves relate directly to sorption isotherms, and directly to the calibration curves of other types of sensors if the transduction mechanisms of those sensors are linear with sorbed concentration.

Our measurements indicate that the nanoparticle-based materials examined sorb vapors with partition coefficients that are of the same order of magnitude as the organic polymers used for comparison, assuming material densities estimated as described above. Strictly from the standpoint of sorption, these materials do not appear to have advantages over polymers as sensing layers.

However, since they can be as good as the polymers at vapor sorption, and may have other desirable properties, they represent candidate layers for acoustic wave sensors whose responses are related to the mass of vapor sorbed. The surface acoustic wave (SAW) device is an example of such a sensor.(15,17-22)

Transduction is also important in sensor performance, and these materials can be used as conducting films on chemiresistor sensors. It has been reported that chemiresistor sensors coated with MPN films can offer lower detection limits than polymer-coated vapor sensors,(11,23,24) such as those based on surface acoustic wave devices or those based on chemiresistors coated with carbon black containing polymers.

If MPN-coated chemiresistors do offer better detection limits than polymer-coated SAW vapor sensors, and the MPN layers and polymer layers absorb similar amounts of vapor, then the lower detection limits for MPN-coated chemiresistors must be due to more signal-to-noise per sorbed vapor molecule. This conditional argument stimulates thinking about signal transduction and nanostructure as a focus for improved sensors.

The implications are that nanostructure matters and suitable nanostructures can lead to excellent signal-to-noise per sorbed molecular unit. Tailoring nanostructure for more sensitive transduction represents an area of opportunity for nanoscience in chemical sensing.

Acknowledgements

The authors thank Dr. Scott Elder for the initial syntheses of some of the nanoparticle materials and Alice Dohnalkova for TEM images. JWG thanks Dr. Arthur Snow for informative and motivating discussion of his work on nanoparticle coated sensors at the Naval Research Laboratory in March 2000 as well as helpful discussions subsequently. This work was funded by the U.S. Department of Energy via Laboratory Directed Research and Development funds administered by the Pacific Northwest National Laboratory. The Pacific Northwest National Laboratory is operated for the U.S. DOE by Battelle Memorial Institute

References

1. Brust, M.; Fink, J.; Bethell, D.; Schiffrin, D. J.; Kiely, C. *J. Chem. Soc., Chem. Commun.* **1995**, 1655-6.
2. Brust, M.; Walker, M.; Bethell, D.; Schiffrin, D. J.; Whyman, R. *J. Chem. Soc., Chem. Commun.* **1994**, 801-2.

162

3. Templeton, A. C.; Wuelfing, W. P.; Murray, R. W. *Acc. Chem. Res.* **2000**, *33*, 27-36.
4. Whetten, R. L.; Shafigullin, M. N.; Khoury, J. T.; Schaaff, T. G.; Vezmar, I.; Alvarez, M. M.; Wilkinson, A. *Acc. Chem. Res.* **1999**, *32*, 397-406.
5. Wohltjen, H.; Snow, A. W. *Anal. Chem.* **1998**, *70*, 2856-2859.
6. Snow, A. W.; Wohltjen, H., Materials, Method and Apparatus for Detecting and Monitoring Chemical Species provisional filing date of November 25, 1997, U.S. Patent 6,221,673 April 24, 2001d April 24, 2001.
7. Evans, S. D.; Johnson, S. R.; Cheng, Y. L.; Shen, T. *J. Mater. Chem.* **2000**, *10*, 183-188.
8. Zhang, H. L.; Evans, S. D.; Henderson, J. R.; Miles, R. E.; Shen, T. H. *Nanotechnology* **2002**, *13*, 439-444.
9. Han, L.; Daniel, D. R.; Maye, M. M.; Zhong, C.-J. *Anal. Chem.* **2001**, *73*, 4441-4449.
10. Zamborini, F. P.; Leopold, M. C.; Hicks, J. F.; Kulesza, P. J.; Malik, M. A.; Murray, R. W. *J. Am. Chem. Soc.* **2002**, *124*, 8958-8964.
11. Cai, Q.-Y.; Zellers, E. T. *Anal. Chem.* **2002**, *74*, 3533-3539.
12. Grate, J. W.; Nelson, D. A.; Skaggs, R. *Anal. Chem.* **2003**, *75*, 1868-1879.
13. Grate, J. W.; Kaganove, S. N.; Bhethanabotla, V. R. *Faraday Discuss.* **1997**, *107*, 259-283.
14. Abraham, M. H.; Andonian-Haftvan, J.; Du, C. M.; Diart, V.; Whiting, G. S.; Grate, J. W.; McGill, R. A. *J. Chem. Soc., Perkin Trans. 2* **1995**, 369-78.
15. Grate, J. W.; Abraham, M. H.; McGill, R. A. *Handb. Biosens. Electron. Noses* **1997**, 593-612.
16. Grate, J. W.; Kaganove, S. N.; Bhethanabotla, V. R. *Anal. Chem.* **1998**, *70*, 199-203.
17. Grate, J. W. *Chem. Rev.* **2000**, *100*, 2627-2647.
18. Grate, J. W.; Kaganove, S. N.; Nelson, D. A. *Chem. Innovation* **2000**, *30*, 29-37.
19. Grate, J. W.; Frye, G. C. *Sens. Update* **1996**, *2*, 37-83.
20. Grate, J. W.; Martin, S. J.; White, R. M. *Anal. Chem.* **1993**, *65*, 987A-996A.
21. Grate, J. W.; Martin, S. J.; White, R. M. *Anal. Chem.* **1993**, *65*, 940A.
22. Grate, J. W.; Abraham, M. H. *Sens. Actuators, B* **1991**, *B3*, 85-111.
23. Snow, A. W., Naval Research Laboratory, personal communication.
24. Comparisons between chemiresistor vapor sensors coated with monolayer protected nanoparticles and surface acoustic wave vapor sensors coated with polymers have been presented by Snow and coworkers at the April 2001 ACS Meeting Defense Applications of Nanomaterials. in San Diego and at the AVS Topical Conference on Understanding and Operating in Threat Environments, Monterey, California, April 30 - May 2, 2002.

Chapter 20

Semiconductor Nanostructures for Detection and Degradation of Low-Level Organic Contaminants from Water

Prashant V. Kamat, Rebecca Huehn, and Roxana Nicolaescu

Notre Dame Radiation Laboratory, University of Notre Dame, South Bend, IN 46556–0579

Introduction

In this new millennium, we are faced with the challenge of cleaning our natural water and air resources. While we enjoy the comforts and benefits that chemistry has provided us, from composites to computer chips, from drugs to dyes, we are faced with the task of treating wastes generated during manufacturing processes and the proper disposal of various products and byproducts. Nanotechnology can provide us with ways to purify the air and water resources by utilizing semiconductor nanoparticles as catalysts and/or sensing systems.

A variety of sensors for environmental applications have been developed in recent years, including SnO_2 based semiconductor systems that have been used as conductometric gas sensors [1,2] and a TiO_2 electrode for determining the chemical oxygen demand (COD) of water [3]. Semiconductor nanostructures have drawn attention in recent years for their potential as chemical sensors. Their exploitation as chemical sensors was prompted by the discovery of photoluminescence from porous-Si and subsequent observation that many organic and inorganic molecules quench this emission [4,5]. The photoluminescence in porous-Si arises from charge carrier recombination in the quantum confined nanosized environments [6,7]. Both energy- and electron transfer mechanisms have been proposed to explain the quenching by organic molecules. Sailor et al have attributed the quenching of the PL by aromatic molecules such as anthracene, pyrene, benzanthracene etc. to energy transfer to the triplet state of the organic substrate [8]. El-Shall and coworkers [9] have shown that the emission from silicon nanocrystals is quenched by aromatic nitrocompounds via electron transfer from the conduction band of the silicon nanocrystal to the vacant orbitals of the quenchers.

The feasibility of using porous silicon nanocrystals as sensors has been demonstrated for detecting nerve gas agents *(10)* and nitroorganics *(9,11)*. Using detection based on electrical conductivity, *(12)* photoluminescence, *(10,11)* or interferometry, *(10)* researchers have achieved sensitivity in the range of ppb-ppm levels.

A desirable feature for the detoxification of air and water is to develop methodologies that can simultaneously sense and destroy toxic chemicals. Such a catalyst system is especially useful when it triggers the photocatalytic operation on demand, i.e., photocatalysis becomes operational only when the system senses the presence of an aromatic compound in the immediate surrounding. Semiconductor nanoclusters such as ZnO and CdSe have an added advantage that they emit quite strongly in the visible region. Size-dependent emission properties of these semiconductor systems have been explored in earlier photochemical studies *(13-17)*. The size dependence properties of these nanoparticles and the ability to tailor the sizes of the nanoassemblies finds ready application in chemical sensors. The thermal conductivity of metal oxide semiconductors is often used to sense organic vapors and gases *(18)*. The visible emission of nanosized ZnO, which usually arises from anionic vacancies, is very sensitive to hole scavengers *(17)*.

Materials and Methods

A small aliquot (usually 250-500 μL) of ZnO slurry was spread over an area of 2 cm^2 of a glass slide and air dried. The semiconductor coated glass plates were then annealed in the oven at 673K for 1 hour. The immobilized semiconductor was inserted in a specially designed flow cell that can be accommodated in a SLM 8000 spectrofluorimeter or on the optical bench of the photolysis set up. Photocatalytic degradation experiments were conducted by illuminating the ZnO film with collimated light from a 150 W Xenon lamp filtered through CuSO$_4$ solution (λ>300 nm).

Results and Discussion

ZnO promises to be a promising candidate for such a specialized environmental application (Figure 1) since it is very sensitive to hole

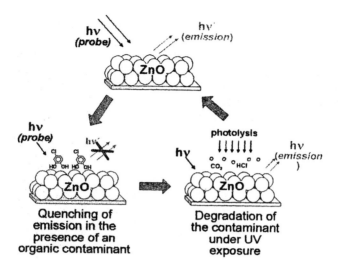

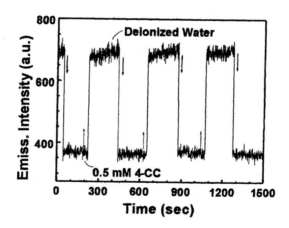

Figure 1. Principle of simultaneous detection and degradation approach in photocatalysis. The emission response of ZnO film to the presence of 4-chlorocatechol is shown on the bottom. (Reproduced with permission from reference 19. Copyright 2002.)

scavengers. The emission is quantitatively quenched by hole scavengers such as iodide ions or phenols *(20)*. The feasibility of employing nanostructured ZnO films for simultaneous sensing and degradation of organic contaminants has been demonstrated recently in our laboratory *(19)*. The ZnO emission is very sensitive to the presence of aromatic compounds such as chlorinated phenols in water. The observed emission quenching is quantitative as the extent of quenching by the organics is determined by the adsorption equilibrium. A detection sensitivity of the order of 1 ppm can be achieved in these systems. A major feature of the ZnO semiconductor system is its ability to degrade the organic contaminant under high intensity UV-illumination.

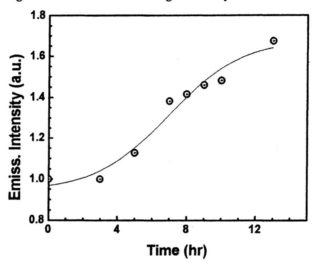

Figure 2. Emission recovery during the degradation of 4-chlorophenol at an UV-irradiated ZnO film. (Reproduced with permission from reference 19. Copyright 2002.)

Experiments illustrating the emission increase during photocatalytic degradation (Figure 2) show that the emission recovery parallels the degradation of aromatic intermediates from the aqueous solution. Thus, emission from ZnO films can be used to monitor the course of a photocatalytic reaction. Though not selective, ZnO emission quenching serves as a probe to sense the presence of aromatic intermediates in water. ZnO film-based nanosensors should be useful in applications such as monitoring the quality of drinking water, or assessing the contamination in underground water. Design engineering of fiber optic probes can miniaturize the sensor design for environmental applications.

Semiconductor nanostructures can play an important role in developing smart materials that can simultaneously sense and destroy harmful chemical contaminants from the environment. Such an application seems to be important as the concern over chemical contamination of drinking water and air needs to be addressed. Application of semiconductor nanoparticles as photocatalysts is still limited by the fact that they respond only to UV-excitation. Continued efforts to extend the response to the visible range have met with limited success. Semiconductor-metal nanocomposites that improve the selectivity and efficiency of the photocatalytic process are expected to draw the attention of future research.

Acknowledgment

This work is supported by the U.S. Department of Energy (DOE), Office of Science, Basic Energy Sciences. This is contribution NDRL No. 4475 from the Notre Dame Radiation Laboratory.

References

1. Sberveglieri, G., *Sensors and Actuators B-Chemical*, **1995**, 23, 103-109.
2. Barsan, N.; Schweizer-Berberich, M.; Gopel, W. *Fresenius Journal of Analytical Chemistry*, **1999**, *365*, 287-304.
3. Kim, Y.C.; Sasaki, S.; Yano, K.; Ikebukuro, K.; Hashimoto, K.; Karube, I. *Analytica Chimica Acta*, **2001**, *432*, 59-66.
4. Brus, L. *Phys. Rev. B*, **1996**, *53*, 4649-4656.
5. Sailor, M.J.; Heinrich, J.L.; Lauerhaas, J.M. *Luminescent porous silicon: Synthesis, Chemistry and applications*, in *Semiconductor Nanoclusters - Physical, Chemical and Catalytic Aspects.*, Kamat, P.V.; Meisel, D. Editors. 1997, Elsevier Science: Amsterdam. p. 209-235.
6. Brus, L. *J. Phys. Chem.*, **1994**, *98*, 3575-3581.
7. Wilson, W.L.; Szajowski, P.F.; Brus, L.E. *Science*, **1993**, *262*, 1242-1244.
8. Song, J.H.; Sailor, M.J. *J. Am. Chem. Soc.*, **1997**, *119*, 7381-7385.
9. Germanenko, I.N.; Li, S.T.; El-Shall, M.S. *J. Phys. Chem. B*, **2001**, *105*, 59-66.
10. Sohn, H.; Letant, S.; Sailor M. J.; Trogler, W.C. *Detection J. Am. Chem. Soc.*, **2000**, *122*, 5399-5400.
11. Content, S.; Trogler, W.C.; Sailor, M.J. *Chemistry, Euro. J.*, **2000**, *6*, 2205-2213.

12. Benchorin, M.; Kux, A.; Schechter, I. *Appl. Phys. Lett.*, **1994**, *64*, 481-483.
13. Empedocles, S.; Bawendi, M. *Acc. Chem. Res.*, **1999**, *32*, 389-396.
14. Li, L.; Hu, J.; Yang, W.; Alivisatos, A.P. *Nano Lett.*, **2001**, *1*, 349-351.
15. Koch, U.; Fojtik, A.; Weller, H.; Henglein, A. *Chem. Phys. Lett.*, **1985**, *122*, 507-10.
16. Bahneman, D.W.; Kormann, C.; Hoffmann, M.R. *J. Phys. Chem.*, **1987**, *91*, 3789-3798.
17. Kamat, P.V.; Patrick, B. *J. Phys. Chem.*, **1992**, *96*, 6829-34.
18. Varfolomeev, A.E., Eryshkin, A.V.; Malyshev, V.V.; Razumov, A. S.; Yakimov, S.S. *J. Anal. Chem.*, **1997**, *52*, 56-58.
19. Kamat, P.V.; Heuhn, R.; Nicolaescu, R. *J. Phys. Chem. B*, **2002**, *106*, 788-794.
20. Kamat, P.V.; Patrick, B. *Photochemistry and photophysics of ZnO colloids.* in *Symp. Electron. Ionic Prop. Silver Halides*; Springfield, Va: The Society for Imaging Science and Technology, 1991.

Chapter 21

Luminescent Silole Nanoparticles as Chemical Sensors for Carcinogenic Chromium(VI) and Arsenic(V)

Sarah J. Toal and William C. Trogler*

University of California at San Diego, 9500 Gilman Drive,
La Jolla CA 92093–0358
*Corresponding author: wct@ucsd.edu

Introduction

The principle objective of the research described herein is to synthesize new materials for the selective sensing of carcinogenic chromium(VI) and arsenic(V) containing species. Both metal ions have been proven to have serious adverse health effects and are regulated by the EPA. As it exists in its predominant environmental form, CrO_4^{2-} is isostructural with the sulfate ion (SO_4^{2-}); hence, sulfate transport proteins uptake CrO_4^{2-}, wherein it damages DNA.[i] Likewise, the predominant environmental form of As(V), AsO_4^{3-}, is isoelectronic with PO_4^{3-}, and so it competes in cellular uptake of phosphate.[ii] The EPA has set maximum contaminant level goals (MCLG) for chromium and arsenic at 0.10 and 0.05 ppm respectively.[iii] The enforcement of these regulations requires the development of sensitive and selective sensors for these carcinogens.

Both chromate and arsenate are oxidizing agents, a property which may be exploited to effect their detection. It has recently been shown that silole containing polymers may be used as sensors specific for TNT and other nitroaromatic oxidant species.[iv] The method of detection is based on luminescence quenching of the silole polymer by the electron accepting analyte. Polysilole luminescence occurs from a LUMO stabilized by σ^*-π^* conjugation arising from the interaction between the σ^* orbital of the silicon chain and the π^* orbital of the butadiene moiety. Recently, it was reported that formation of methyl(phenyl)tetraphenylsilole colloids leads to a significant increase in luminescence.[v] Although poly(tetraphenyl)silole is insensitive to simple inorganic oxidants, functionalization of the silole unit with hydrogen bonding substituents has been performed to incorporate binding regions for anionic oxidants, such as chromate and arsenate. Nanoparticle colloids of poly(tetra-phenyl)silole show not only increased luminescence, but also much greater sensitivity in analyte detection.

Materials and Methods

The silole compound used for the detection studies **1A**, 1-(3-aminopropyl)-1-methyl-2,3,4,5-tetraphenylsilole, was designed to have an amino terminated alkyl substituent on the silicon atom, which hydrogen bonds to the oxygens of the chromate and arsenate analytes in solution. It was synthesized from the chloroplatinic acid (H_2PtCl_6) catalyzed hydrosilation of allylamine by 1-methyl-2,3,4,5-tetraphenylsilole. The latter silole was prepared by a modified literature procedure.[vi]

Results and Discussion

The UV-vis absorption spectrum of silole **1A** (Figure 1) exhibits two absorption bands at 248, and 360 nm. The longest wavelength absorption is typical of silole compounds,[5] and is attributed to the π-π* transition of the metallole ring.

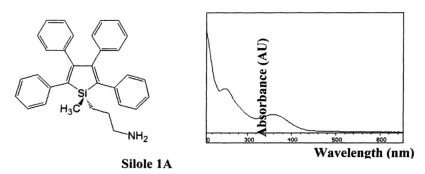

Silole 1A

Figure 1: **UV-vis data for 1A in THF**

Fluorescence studies were performed with an excitation wavelength of 360 nm on a 10 μm solution of **1A**. Silole **1A** is a weak luminophore in THF solution with a λ_{max} = 475 nm. However, addition of water to the THF solution causes precipitation as a colloid and increases the luminescence dramatically. Atomic Force Microscopy images were taken on a glass coverslip coated with the nanoparticles, and they reveal the size of the nanoparticles to be on the order of 80-100 nm (Figure 2a). Surface coating of the nanoparticles with SiO_2 by treatment with $Si(OEt)_4$ produces particles with sizes on the order of 150 nm, and slows the rate of nanoaggregate settling.

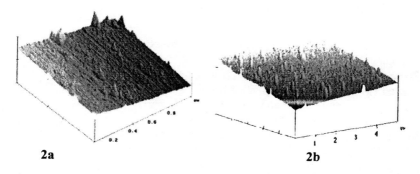

2a   **2b**

Figure 2: **AFM images of silole 1A nanoparticles: 2a, uncoated (1 μm x 1 μm scan). and 2b. silicate coated (5 um x 5 um scan)**

For uncoated nanoparticles, chromate and arsenate detection studies were made on a 10 μm solution of **1A** in 1:9 THF:H₂O. Luminescence quenching is observed at concentrations as low as 0.50 and 5 ppm of Cr(VI) and As(V) respectively. Stern-Volmer plots for the quenching, however, are non-linear, which may reflect nanoparticle surface saturation (Figures 3a-b).

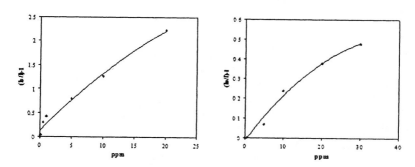

Figure 3: **Stern-Volmer plots for luminescence quenching by 3a) Cr(VI) and, 3b) As(V)**

It has been shown that silole compounds functionalized with anion binding sites show a capability of detecting chromate and arsenate at low concentrations.

This research is funded by the U.S. EPA-Science to Achieve Results (STAR) program grant # R829619.

References

[i] Lippard, S.J.; Berg, J.M. *Principles of Bioinorganic Chemistry;* University Science Books; Mill Valley, CA, 1994

[ii] Kaim, W.; Schwederski, B.; *Bioinorganic Chemistry: Inorganic Elements in the Chemistry of Life;* John Wiley & Sons, Inc., New York, NY, 1994

[iii]http://www.epa.gov/safewater/dwh/c-ioc/chromium.html; http://www.epa.gov/safewater/arsenic.html

[iv] a) Sohn, H.; Calhoun, R. M.; Sailor, M.J.; Trogler, W.C. *Angew. Chem. Int. Ed.* **2001**, 40, 2104; b) Sohn, H; Sailor, M.J.; Magde, D.; Trogler, W.C.; *J. Am. Chem. Soc.* **2003**

[v] Luo, J. et. al., *Chem. Commun.*, **2001**, 1740

[vi] Boudjouk, P.; Sooriyakumaran, R.; Han, B. *J. Org. Chem.,***1986**, 51, 2818

Chapter 22

A Nanocontact Sensor for Heavy Metal Ion Detections

V. Rajagopalan, S. Boussaad, and N. J. Tao

Department of Electrical Engineering and Center for Solid State
Electronics, Arizona State University, Tempe, AZ 85287

Introduction

The ability to detect trace amount of metal ions is important because of the toxicity of heavy metal ions on a broad range of living organisms and the consequence of heavy metals not being biodegradable. To date, heavy metals in environment are usually measured with spectroscopic techniques, which require samples to be collected and transported to the laboratory for analysis. *In situ* measurements are highly desirable because they provide an early detection of trace metal contaminants while minimizing errors, labor and cost associated with collection, transport, and storage of samples for subsequent laboratory analysis. One of the promising *in situ* sensors is based on anodic stripping voltammetry.[1-5] To achieve detection limits of trace elements in the μg/L to ng/L range, a preconcentration or separation technique is normally required, in which metal ions are electrochemically deposited onto a Hg electrode. Following the preconcentration, the electrode potential is scanned to a more positive value to oxidize and strip the deposited metals from the electrode. The oxidation current reaches a peak at a potential characteristic of the metal, which provides the identity of the metal. Other methods include fluorescence-detection,[6-12] porphyrin derivatives as complexing agent,[7] array-based sensors,[13, 14] magnetic effects,[15] ion-sensitive field effect transistors[16] and microfabricated cantilevers.[17, 18] While each method has its own advantages and disadvantages, a simple and inexpensive yet reliable and sensitive sensor requires further effort.

In the present work, we demonstrate a method to detect trace amount of metal ions using such electrodes. Our method starts with electrodes separated with a narrow gap (Fig. 1). When the electrodes are exposed to a solution containing heavy metal ions, the ions can be deposited into the gap by controlling the electrode potentials. Once the deposited metal bridges the gap, a sudden jump in the conductance between the electrodes occurs, which can be easily detected. Since the gap can be made as narrow as a few nm or less, the deposition of even a few ions into the gap is enough to trigger a large change in the conductance, thus providing a sensitive detection of metal ions. Our previous experiments (19, 20) have shown that the conductance of such a small bridge is quantized and given by NG_0, where N is an integer and G_0 is conductance quantum (=$2e^2/h$, e is the electron charge and h is the Planck constant). For this reason, the bridge is often called quantum point contact. (21-24)The metal bridge can also be stripped off (or dissolved) by sweeping the potential anodically. The potentials at which deposition and dissolution take place provide identity of the metal ions, a principle similar to that of anodic stripping analysis.

Materials/Methods

We started with a pair of Au electrodes separated with a μm-scale gap on a Si or glass substrate in 0.2 M HCl. (20, 25) We then applied a bias voltage (~ 1.2 V) between one electrode and an external resistor connected in series with the other electrode. Initially, the entire bias voltage fell across the gap between the two electrodes, consequently, Au atoms were etched away from the anode and deposited on the cathode. Because the deposition was a diffusion limited process, the cathode grew directionally towards the anode. When the gap between the cathode and anode decreased below a few nm, tunneling current began to flow across the gap and the effective resistance of the gap dropped sharply. As a result, the voltage across the gap also dropped sharply and eventually determined the etching and deposition process and left us with a pair of electrodes separated with a gap of 0.4 – 0.6 nm wide (estimated from the tunneling current) (Inset of Fig. 1). The success rate to form such a gap was about 50% (several hundred of electrodes pairs), and the rest ended up with point contacts between the two electrodes. However, many of the contacts could be re-broken by flowing a large current through them based on electromigration effect, (26) which allowed us to re-start the etching and deposition procedure described above.

For heavy metal ion detection, we used a bipotentiostat (Pine Instruments, Model AFRDE 5) to control the potentials of the two electrodes and determine the conductance between the two electrodes at the same time. For potential control, we used a Ag wire as quasi-reference electrode and a Pt wire as the counter electrode. We tested two metal ions, Pb^{2+} and Cu^{2+}, by dissolving $Pb(NO_3)_2$ and $CuSO_4$ in 50 mM HNO_3 and H_2SO_4 supporting electrolytes, respectively. The metal ions are deposited into the gap of the electrodes either by holding the potentials of the two electrodes close to the reduction potentials of the metals or by scanning the potentials between a preset upper and lower limits. In the later case, we can observe deposition & dissolution of the metal ions reversibly. Once the deposited metal bridges the gap, the conductance between the two electrodes increased abruptly, which was measured with the bipotentiostat and recorded with a digital oscilloscope.

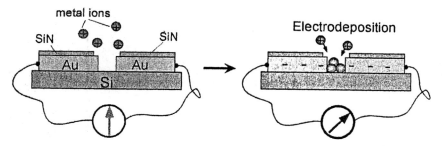

Figure 1. (A) A drop of sample solution is placed onto a pair of nanoelectrodes separated with an atomic scale gap on a silicon chip. (B) Holding the nanoelectrodes at a negative potential, electrochemical deposition of a single or a few metal atoms into the gap can form a nanocontact between the two nanoelectrodes and result in a quantum jump in the conductance.

Results and Discussions

After forming the electrodes, we held the electrodes potential below and above the electrodeposition potential (reduction). When the potentials were held above the electrodeposition potentials, no conductance change was observed. However, below the electrodeposition potentials, conductance jumped above $1G_0$, the conductance of a single atom contact between the two electrodes. Figure 2a shows an example of the experiment of Cu^{2+}. The conductance is not measurable initially, but it jumped above $1G_0$ several seconds after holding the potential   below the Cu deposition potential. The conductance continued to increase as the electrodeposition continued at the potential. The amount of time that takes to form the contact depends on the concentration of the ions. We have determined the deposition as a function of concentration and plotted the results in Fig. 2b. As we might expect, the time increases as the concentration decreases, but it is almost proportional to the logarithmic of the ion concentration. A qualitative understanding the time dependence on the concentration requires a good knowledge of the electrode geometry, which will be studied in the future. Nonetheless, the observed simple dependence of the deposition time on the concentration provides us with a method to estimate the concentration of ions. This measurement also shows that we can detect a concentration of 10 nM Cu^{2+}. We performed the similar experiments for Pb^{2+} ions, whose electrodeposition potential is lower than that of Cu^{2+}.

176

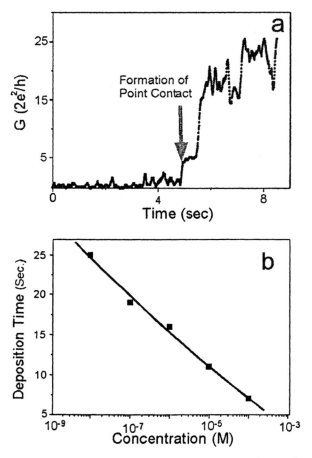

Figure 2 (a) Conductance of a pair of Au electrodes in 10^{-4} M Cu^{2+} solution as a function of time. The electrodes were held at the electrodeposition potential of Cu^{2+} with a bias of 50 mV applied between the two electrodes. The sudden jump in the conductance (marked by an arrow) corresponds to the formation of a Cu point contact between the Au electrodes. (b) Deposition time required to form a Cu point contact vs. ionic concentration, where the solid line is a guide to eye.

In summary, we have demonstrated a simple method to detect trace amount of metal ions (0.1 nM) using a pair of electrodes separated with atomic-scale gap. By controlling the potentials of the electrodes, metal ions in solution can be deposited in the gap and form a point contact between the two electrodes, which can be sensitively detected via the quantum jump in the measured conductance. The deposition time that it takes for the formation of a point contact increases as the ionic concentration decreases. We have also shown that different ions form point contacts or dissolve point contacts at different potentials, which can be used to identify the ionic species.

Acknowledgment

We thank Prof. F.M. Chou and Y. Cai for valuable comments and EPA(R82962301) for financial support.

References

1. Wang, J. *Stripping analysis: Principles, Instrumentation and Applications*; VCH Publishers: Deer Field Beach, Florida, 1985.
2. Zhou, F. M.; Aronson, J. T.; Ruegnitz, M. W. *Anal. Chem.* **1997**, *69*, 728-733.
3. Lai, R.; Huang, E. L.; Zhou, F. M.; Wipf, D. O. *Electroanalysis* **1998**, *10*, 926-930.
4. Herdan, J.; Feeney, R.; Kounaves, S. P.; Flannery, A. F.; Storment, C. W.; Kovacs, G. T. A.; Darling, R. B. *Environ. Sci. Technol.* **1998**, *32*, 131-136.
5. Feeney, R.; Kounaves, S. P. *Anal. Chem.* **2000**, *72*, 2222-2228.
6. Segura-Carretero, A.; Costa-Fernandez, J. M.; Pereiro, R.; Sanz-Medel, A. *Talanta* **1999**, *49*, 907-913.
7. Delmarre, D.; Meallet, R.; Bied-Charreton, C.; Pansu, R. B. *Journal of Photochemistry and Photobiology a-Chemistry* **1999**, *124*, 23-28.
8. Kuswandi, B.; Narayanaswamy, R. *Analytical Letters* **1999**, *32*, 649-664.
9. Fabbrizzi, L.; Licchelli, M.; Parodi, L.; Poggi, A.; Taglietti, A. *European Journal of Inorganic Chemistry* **1999**, 35-39.
10. Malcik, N.; Caglar, P. *Sensors and Actuators B-Chemical* **1997**, *39*, 386-389.
11. DeSantis, G.; Fabbrizzi, L.; Licchelli, M.; Mangano, C.; Sacchi, D.; Sardone, N. *Inorganica Chimica Acta* **1997**, *257*, 69-76.
12. Czolk, R.; Reichert, J.; Ache, H. J. *Sensors and Actuators B-Chemical* **1992**, *7*, 540-543.
13. Prestel, H.; Gahr, A.; Niessner, R. *Fresenius Journal of Analytical Chemistry* **2000**, *368*, 182-191.

14. Kim, J.; Wu, X. Q.; Herman, M. R.; Dordick, J. S. *Analytica Chimica Acta* **1998**, *370*, 251-258.

15. Petrov, N. K.; Kuhnle, W.; Fiebig, T.; Staerk, H. *Journal of Physical Chemistry a* **1997**, *101*, 7043-7046.

16. Ben Ali, M.; Kalfat, R.; Sfihi, H.; Chovelon, J. M.; Ben Ouada, H.; Jaffrezic-Renault, N. *Sensors and Actuators B-Chemical* **2000**, *62*, 233-237.

17. Ji, H. F.; Finot, E.; Dabestani, R.; Thundat, T.; Brown, G. M.; Britt, P. F. *Chem. Comm.* **2000**, *6*, 457-458.

18. Ji, H. F.; Thundat, T.; Dabestani, R.; Brown, G. M.; Britt, P. F.; Bonnesen, P. V. *Anal. Chem.* **2001**, *73*, 1572-1576.

19. Li, C. Z.; Tao, N. J. *Appl. Phys. Lett.* **1998**, *72*, 894-897.

20. Li, C. Z.; Bogozi, A.; Huang, W.; Tao, N. J. *Nanotechnology* **1999**, *10*, 221-223.

21. Krans, J. M.; Ruitenbeek, J. M. v.; Fisun, V. V.; Yanson, I. K.; Jongh, L. J. d. *Nature* **1995**, *375*, 767-769.

22. Muller, C. J.; Krans, J. M.; Todorv, T. N.; Reed, M. A. *Phys. Rev. B* **1996**, *53*, 1022-1025.

23. Landman, U.; Luedtke, W. D.; Salisbury, B. E.; Whetten, R. L. *Phys. Rev. Lett.* **1996**, *77*, 1362-1365.

24. Pascual, J. I.; Mendez, J.; Gomez-Herrero, J.; Baro, A. M.; Garcia, N.; Binh, V. T. *Phys. Rev. Lett.* **1993**, *71*, 1852-1855.

25. Li, C. Z.; He, H. X.; Tao, N. J. *Appl. Phys. Lett.* **2000**, *77*, 3995-3997.

26. Park, H.; Lim, A. K. L.; Alivisatos, A. P.; Park, J.; McEuen, P. L. *Appl. Phys. Lett.* **1999**, *75*, 301-303.

Chapter 23

Ultrasensitive Pathogen Quantification in Drinking Water Using Highly Piezoelectric Microcantilevers

Wan Y. Shih[1], G. Campbell[2], J. W. Yi[1], H. Luo[1], R. Mutharasan[2], and Wei-Heng Shih[1]

Departments of [1]Materials Engineering and [2]Chemical Engineering, Drexel University, Philadelphia, PA 10104

Current development of pathogen detection in water relies on filtration culture methods[i,ii] and fluorescence-based methods,[iii-v] e.g., fluorescence probes methods[iv] and DNA microarray methods.[v] These techniques, however, do not lend themselves for in-situ, rapid, quantitative measurements. With the filtration culture methods, sample water is passed through a filter that is pretreated for visualization of the target pathogen. Growth of colonies on the filter indicates the presence of the target pathogen in the test water. Both the fluorescently labeled probe methods[iv] and the DNA microarray methods[v] rely on detection using fluorescence spectroscopy, which is not quantitative. There is an immediate need for rapid, quantitative, and specific pathogen detection to ensure the safety of natural and manmade water supplies, including source, treated, distributed and recreational waters.

Another development of biosensing technologies relies on silicon-based microcantilevers [vi-x] due to their availability and ease of integration with existing silicon based technologies. All silicon-based microcantilevers rely on external optical components for deflection detection. Antibody receptors were coated on the surface of the microcantilevers to bind target DNA, protein molecules, or bacteria.[vi-x] The adsorbed target molecules can be detected by monitoring the mechanical resonance frequency of the microcantilever. The adsorption of target molecules causes a change in the microcantilever's mass, which in turn causes a shift in the resonance frequency. Because of the small sizes of the silicon-based microcantilevers, about 100 μm in length, they exhibit high mass-detection sensitivity. The mass change per Hz is about $\Delta m/\Delta f \sim 10^{-12}$ g/Hz,[xi,xii] where Δm and Δf respectively denote the mass change and the corresponding resonance frequency change due to the binding of the target molecules. However, the required optical components are large and complex, requiring precise alignment. Moreover, immersing the silicon-based microcantilevers in water reduces the resonance intensity by an order of magnitude,[xiii] reducing the Q factor, defined as the ratio of the resonance peak frequency relative to the resonance peak width at half peak height, to about one, thus prohibiting the use of silicon-based microcantilever for in-water detection. The main reason that silicon based microcantilevers cannot have high resonance signal in water is that the microcantilevers are not piezoelectric.

We have demonstrated highly piezoelectric lead zirconate titanate/stainless steel cantilevers for mass detection,[xiv] liquid density and viscosity sensing,[xv] yeast cell quantification,[xvi] protein detection,[xvii] and protein-antibody specific binding detection.[xviii] Moreover, we showed that unlike a silicon-based microcantilever, a piezoelectric cantilever can still maintains a high Q value in water, making it particularly suitable for in-water detection. As an example, we showed the in-air and in-water resonance spectra of the PZT/stainless steel cantilever that was later used for yeast detection in Fig. 1(a). Detection of yeast cells is shown in Fig. 1(b) where the resonance frequency shift versus time of the cantilever immersed in different yeast concentrations is shown. Note that at different yeast concentrations, the resonance frequency shift rises differently with time. The kinetics of the resonance frequency shift bears the information of the yeast concentration and can be yeast concentration quantification.[16]

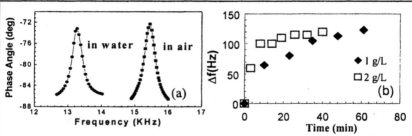

Fig. 1: (a) First-mode resonance frequency spectra of the yeast detecting cantilever, (b) Resonance frequency shift versus time after the cantilever was immersed in the 1g/L (full diamonds) and 2 g/L (open squares) suspension.

Currently we are extending this approach to detect E. coli 0157:H7 in water using a piezoelectric PZT/stainless steel cantilever. The cantilever tip was coated with the antibody of E. coli 0157:H7. The detection of the E. coli 0157:H7 is shown in Fig. 2(a) where the resonance frequency shift with time due to the binding of the E coli cells to the cantilever tip. The detachment of the E coli 0157:H7 cells from the cantilever tip at pH = 2 in a glycine buffer is shown in

Fig. 2(b). A scanning electron micrograph of the E coli 0157:H7 cells is shown in the insert of Fig. 2(a).

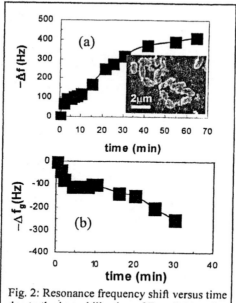

Fig. 2: Resonance frequency shift versus time due to the immobilization of E. coli (a) and detachment of E coli (b). The insert shows a SEM micrograph of immobilized E. coli.

Examining the dependence of of the mass detection sensitivity, $\Delta f / \Delta m$, on the length and width of the cantilever and the resonance mode, we showed that the mass detection sensitivity of the cantilever increases with a decreasing cantilever size (see Fig. 3) as $\Delta f / \Delta m \propto 1/wL_p^3$, where L_p is the length, w the width, signifying that a piezoelectric cantilever of 50 um in length would reach a detection sensitivity of 10^{-14} g/Hz, smaller than the mass of a single bacterium. This offers the potential for unprecedented detection sensitivities in ultra low concentrations.

To achieve 10^{-14} g/Hz detection sensitivity, currently, we are making highly piezoelectric microcantilevers by integrating highly piezoelectric lead magnesium niobate-lead titanate (PMN-PT) thick layer in the microfabrication process. We have succeeded in making PMN-PT thick layers using a novel nano-layer coating approach and reasonably thick (e.g., 1-10 μm) layers with excellent piezoelectric properties were made using a modified sol-gel approach. The X-ray diffraction pattern and a SEM micrograph of the PMN-PT thick layer on a titanium foil are shown in Fig. 4(a) and (b).

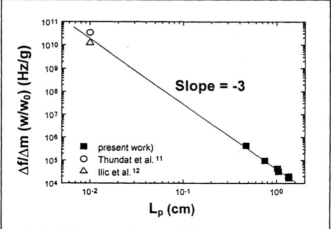

Fig.3: (a) Resonance frequency shift per unit mass change with a normalized width versus cantilever length.

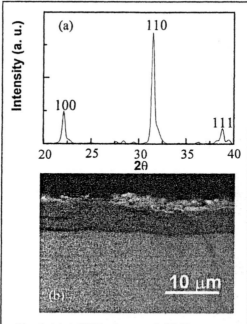

Fig. 4: (a) A SEM micrograph (b) X-ray diffraction pattern of a 6 layer PMN-PT thick film on titanium sintered at 700°C for 2 hours. The film has a thickness of about 6 μm.

Table I: Comparison of the proposed highly piezoelectric microcantilever pathogen detector with current technologies

	Detection method	specificity	sensitivity in air	sensitivity in water	quantitative	Potential for ultra low concentration detection
Quartz crystal microbalance/antibody	electrical	yes	10^{-8} g/Hz	ineffective	yes	no
Silicon based cantilevers/antibody	optical	yes	10^{-12} g/Hz	ineffective	yes	no
Fluorescence labeled probes	optical	yes			no	no
DNA microarrays	optical	no			no	no
Filter culture methods	Visual/optical	yes			no	no
Proposed 50 μm long piezoelectric microcantilevers/antibody	electrical	yes	10^{-14} g/Hz	10^{-14} g/Hz	yes	yes

It is expected that highly piezoelectric PMN-PT microcantilevers of 50 μm in length that offer unprecedented detection sensitivity will emerge from the present study. In Table I, we summarize the pros and cons of currently available biodetection technologies. Clearly, applied to detect pathogens in drinking water, each technique has its shortcomings. The present approach of using highly piezoelectric microcantilevers coupled with antibodies specific to target microbes immobilized at the cantilever tip has the right attributes for *direct, ultrasensitive, rapid, label-free, in-situ, and specific quantification* of pathogens in drinking water as indicated by our preliminary results shown above.

Acknowledgements

The work is supported in part by the Environmental Protection Agency (EPA) under Grant No. R829604, the National Science Foundation under Grant No. BES–0120321, and the National Aeronautics and Space Administration (NASA) under Grant No. NAG2-1475.

References

[i] KP Brenner, CC Rankin, YR Roybal, GN Stelma Jr, PV Scarpino, and AP Dufour, *Appl. Environ. Microbiol.* **59**, 3534 (1993).

[ii] EPA online publication, http://www.epa.gov/nerlcwww/online.htm

[iii] S. Nie and R. N. Zare, "Optical Detection of Single Molecules," Annu. Rev. Biophys. Biomol. Struct., 26, 567-96 (1997).

[iv] M. B. Tabacco, EPA contract number: 68D01016, "A new Biosensor for Rapid Identification of Bacteria Pathogens."

[v] Darrell P. Chandler and Ricardo De Leon, EPA Grant No. R828039, "Detection of Emerging Microbial Contaminants in Source and finished drinking water with DNA microarrays."

[vi] J. Fritz, M. K. Baller, H. P. Lang, H. Rothuizen, P. Vettiger, E. Meyer, H.-J. Guntherodt, Ch. Gerber, J. K. Gimzewski, *Science*, **288**(5464), 316-8 (2000).

[vii] A. Schemmel and H. E. Gaub, *Rev. Sci. Instrum.* **70**(2), 1313-7 (1999).

[viii] D. R. Baselt, G. U. Lee, R. J. Colton, *J. Vac. Sci. Technol.* **B14**(2), 798-3 (1996).

[ix] G. U. Lee, D. A. Kidwell, R. J. Colton, *U. S. Pat. Appl.* (1996) 38 pp. Avail. NTIS Order No. PAT-APPL-8-505 547.

[x] W. Han, S. M. Lindsay, T. Jing, *Appl. Phys. Lett.* **69**(26), 4111-3 (1996).

[xi] T. Thundat, E. A. Wachter, S. L. Sharp, and R. J. Warmack, *Appl. Phys. Lett.*, **66**, 1695 (1995).

[xii] B. Ilic, D. Czaplewsli, H. G. Craighead, P. Neuzil, C. Campagnolo, and C. Batt, *Appl. Phys. Lett.*, **77**, 450 (2000).

[xiii] P. I. Oden, G. Y. Chen, R. A. Steele, R. J. Warmack, and T. Thundat, Appl. *Phys. Lett.*, **68**, 3814 (1996).

[xiv] G. Campbell, J. W. Yi, R. Muthasan, W. Y. Shih, and W.-H. Shih, "In-Situ Detection of Ecoli Cells Using Piezoelectric PZT/Stainless Steel Cantilevers," *J. Appl. Phys.* to be submitted.

[xv] W. Y. Shih, X. Li, H. Gu, W.-H. Shih, and I. A. Aksay, *J. Appl. Phys.*, **89**, 1497 (2001).

[xvi] J. W. Yi, W. Y. Shih, R. Mutharasan, and W.-H. Shih, "In-Situ Yeast cell Detection Using Piezoelectric PZT/stainless Steel Cantilevers," *J. Appl. Phys.*, **93**(1), 619-625 (2003)

[xvii] J. W. Yi, W. Y. Shih, R. Mutharasan, and W.-H. Shih, "In-Situ Detection of Avidin Immobilization Using Piezoelectric PZT/stainless Steel Cantilevers," *J. Appl. Phys.*, to be submitted.

[xviii] J. W. Yi, W. Y. Shih, R. Mutharasan, and W.-H. Shih, "In-Situ Detection of Biotinylated Polystyrene Spheres Using Piezoelectric PZT/stainless Steel Cantilevers," *J. Appl. Phys.*, submitted.

Chapter 24

Nanostructured Sorbents for Solid Phase Microextraction and Environmental Assay

R. Shane Addleman[1,*], Oleg B. Egorov[1], Matthew O'Hara[1], Thomas S. Zemanian[1], Glen Fryxell[1], and Dean Kuenzi[2]

[1]Pacific Northwest National Laboratory, 902 Battelle Boulevard, P.O. Box 999, Richland, WA 99352
[2]Eastern Oregon State College, La Grande, OR 99352
*Corresponding author: Shane.Addleman@pnl.gov

Introduction

An array of environmental problems necessitates the development of rapid, portable, cost effective measurement systems for determining trace levels of contaminants. Unfortunately instruments robust enough to operate outside of a laboratory typically do not have sufficient sensitivity or selectivity. One approach to improving any analytical method is to separate the analyte(s) from the sample matrix and concentrate them into a smaller volume prior to measurement. When this preconcentration is done with small quantities of solid sorbents the method is typically referred to as solid-phase microextraction (SPME).

SPME bridges the gap between sample collection and sample analysis in a single step. After the analyte is captured in the sorbent it can be transferred to an instrument for analysis or, in some cases, assayed directly. SPME significantly reduces the amount of chemicals and solid waste generated during the assay process. SPME is also typically faster than conventional liquid extraction processes and is very amenable to field applications. SPME is an equilibrium process and the amount of analyte extracted is ultimately limited by the magnitude of the partition coefficient of the analyte between the sample matrix and the sorbent material. Consequently, effective SPME materials require high affinities for the target analyte and large surface areas.[1]

Self-assembled monolayers on mesoporous supports (SAMMS) have a number of properties that make them ideal sorbents for SPME. The mesoporous silica structure is produced through a surfactant templated sol-gel processes[2]

186

which endows SAMMS with high surface areas and rigid open pore structure (thereby providing short diffusional path lengths and rapid sorption kinetics).[3] A monolayer with an appropriate terminal ligand upon the surface of these mesoporous silica structures provides a high density of chemically selective complexation sites.[3,4] Well-established ligand chemistry[5] enables a wide range of monolayer chemistries allowing significant flexibility in the target analyte(s).[6-13] The combination of high density monolayers and the high surface mesoporous structure results in hybrid nanostructured sorbents (SAMMS) ideally suited for SPME ; these materials offer high capacity, rapid sorption kinetics, and interfacial chemistry easily tailored for a wide variety of analytes. SAMMS have been shown to be excellent sorbents for many environmentally relevant targets, including heavy metals such as mercury, lead and cadmium [6-9], chromate and arsenate [10], and radionuclides such as ^{99}Tc, ^{137}Cs[11], uranium and the actinides [13]. Herein we discuss the advantages, challenges and limitations for using SAMMS for SPME and environmental characterization.

Materials and Methods

Mesoporous Supports

There has been a great deal of work in the synthesis of nanostructured materials recently, particularly in the area of surfactant templated synthesis of mesoporous ceramic materials.[2, 14] Synthetic methods have been developed to prepare these materials in a variety of morphologies (lamellar, cubic, hexagonal, etc.) with structural features ranging from about 20Å to as much as 300Å.[14] The micelle shape and structure dictates pore size and structure. The material used for this work is based on commercially available MCM-41mesoporous silica (Mobil). The material had an average pore diameter of approximately 35Å and a surface area of approximately 980 m²/g. This pore diameter is smaller than ideal for monolayer derivatization, but was used in this study as it is readily available. The SAMMS generated using this substrate were evaluated for SPME applications, and found that they performed well as preconcentration media, in spite of the relatively small pore size. Studies with larger pore materials are currently underway.

Self-assembled Monolayers

The self-assembly of a monolayer is the spontaneous aggregation of molecules into an ordered, organized macromolecular array on an interface, yielding a functional layer only a single molecule thick. [15] The self-assembly process is driven by the attractive forces between the molecules themselves (*e.g.* van der Waals' interactions, hydrogen bonding or dipole-dipole interactions), as well as the attractive forces between the molecule and interface (*e.g.* hydrogen bonding, acid/base interactions, *etc.*). It is important to recognize that the molecular spacing in a self-assembled monolayer is dictated by the aggregation efficiency and the molecular "footprint" of the monomer. To effectively exploit the high surface area it is advantageous to install a fully dense self-assembled monolayer across the entire silica surface. Only through the judicious choice of reaction conditions (*i.e.* solvent identity, water concentration, water location, and reaction temperature) can self-assembly efficiently take place, resulting in a dense, uniform coating of the surface. Supercritical fluids, particularly CO_2, are effective reaction media in which to carry out silane self-assembly within mesoporous ceramics. [16] Prudently chosen supercritical fluids effectively solvate the siloxane monomers, assist in the monolayer formation processes, and by virtue of their gas-like viscosities facilitate mass transport of reactants throughout the nanoporous ceramic matrix.

The specific materials, synthetic procedures, and analytical methods utilized in these studies have been previously described in detail elsewhere.[7-13]

Results and Discussions

Heavy Metals; Uptake and Selectivity

Initial efforts into developing SAMMS as sorbents for metal ions focused on the sequestration of soft heavy metals (such as Hg^{2+}, Cd^{2+}, *etc.*) due to their impact on the environment and human health. Since these heavy metals are "soft" Lewis acids, monolayers terminated in a "soft" Lewis base (in particular alkyl thiols) have high affinities for mercury and other heavy metals of environmental concern such as Cd^{2+}, Pb^{2+}, and Tl^{1+}. Consequently, thiol SAMMS were developed and found to be excellent sorbents for the soft heavy metals, exhibiting high partition coefficients (often exceeding 10^6) and high loading capacity (binding more than 500 mg Hg^{2+} per gram of SAMMS).[6-9] The binding affinity of thiol SAMMS for Hg^{2+} is so high that even complexants like EDTA cannot compete. [9] Ubiquitous cations such as Na^{1+}, K^{1+}, Ca^{2+},

Mg^{2+}, Fe^{2+} and Zn^{2+} were found to provide no measurable competition to Hg^{2+} for the thiol binding sites; a result also predicted by hard-soft acid base theory.

The large partition coefficients and selectivity of the thiol SAMMS/Hg^{2+} system enables effective SPME from complex environmental matrices. Due to the high partition coefficients and capacity of the sorbent material, trace analyte uptake is linear over very large preconcentration ranges. Selective SPME preconcentration (up to a factor of 10^6) of heavy metals with thiol SAMMS from actual river and aquifer water has been demonstrated. Sorbent material morphology can also be very important for uptake dynamics. For example, small particulates (< 5 microns) can block flow in packed microfluidic SPME columns. The hydrophobic nature of some monolayers can create difficultly in wetting the functional surfaces of the nanopores and reduce analyte uptake. Humic acids, suspended organics, and colloids in surface and ground waters can significantly reduce analyte uptake by plugging the SPME column. However, these problems can be resolved with intelligent application of techniques such as particle sorting, sequential solvent washing, and in-line filtration methods.

Detection

X-ray fluorescence (XRF) spectroscopy coupled with SAMMS SPME enables rapid detection of trace levels of toxic metals. Metal assay with XRF spectroscopy is based upon the quantitative measurement of characteristic elemental X-rays emitted from an atom after bombardment by higher energy photons. XRF is capable of direct rapid multi-elemental analyses of solids and liquids but has limited (part per million level) sensitivity. SPME with SAMMS provides a mechanism to concentrate the target analyte, thereby significantly lowering detection limits. Concentrating the analyte into SAMMS also mitigates two primary factors limiting XRF sensitivity; the heterogeneity of the sample matrix and the particle sizes. The typically small, and potentially uniform, SAMMS particles (< 200 microns) allow the excitation radiation and subsequent characteristic x-rays to efficiently and uniformly pass through the silica SAMMS structure.

Table 1 shows the concentration of trace metals in an SPME thiol SAMMS column (15 mg) after 200 mL of river water (pH 8) spiked with of 1 ppb of selected metal ions was pumped through at 1 ml/min. The difference in the preconcentration of the metals in the SPME SAMMS column is due to the competitive equilibrium processes between the sample solution and monolayer surface. However, the SAMMS SPME clearly preconcentrated all the metals from a level (1ppb) well below the EPA drinking water limits (listed in the right column) to a range easily measurable by the XRF. Using SAMMS selective for oxoanions we have obtained similar results for chromate and arsenate anions. The actual limit of detection for metals in environmental samples with this approach will depend upon a number of variables including the target metal, the sample matrix, the type of SAMMS, the relative volumes of sample and sorbent, and the sensitivity of the XRF system. However, it is clear that SPME with SAMMS type materials can be a powerful tool for environmental sample preconcentration and assay.

Table 1. Performance of thiol SAMMS and XRF

Metal	SAMMS Concentration[a] (ppm)	XRF LOD[b] (ppm)	EPA DWS[c] (ppm)
Hg^{2+}	7.2	2	0.002
Cu^{2+}	57.5	2	1.3
Pb^{2+}	7	2.4	0.015
Ag^{1+}	16.3	7	0.1
Cd^{2+}	18.7	7	0.005

a 200ml of river water spiked with of 1 ppb of metal ions and passed through a 15 mg SPME thiol SAMMS column
b Limit of detection (LOD) for XRF system utilized
c EPA Drinking water standards (DWS) from www.epa.gov/safewater/mcl.html

Radionuclides

SPME with SAMMS can provide similar advantages for preconcentration of radionuclides. A number of SAMMS sorbent materials have been recently created for the sorption of actinides [12-13]. SAMMS have also been created that are excellent sorbents for the environmentally relevant fission products such as ^{99}Tc, and ^{137}Cs [11]. Radionuclides lend themselves particularly well to direct assay of the SPME sorbent through direct counting of radiation from the SPME material after preconcentration. Coupling SPME SAMMS with nuclear detection systems enhances the selectivity and sensitivity for detection of the selected radionuclides. In addition to concentrating the target radionuclide(s) the SPME sorbent excludes interferents and provide a uniform well defined sample geometry; a critical issue for isotopic assay. Prudently chosen sorbents coupled with selected detection methods can provide rapid assay capability for both field screening and isotopic laboratory measurements. The detection scheme employed depends upon the type of radiation (alpha, beta, gamma, neutron) emitted from the radionuclide and whether or not isotopic information is desired. To date we have found SPME with SAMMS very effective for determination of gamma, beta, and alpha emitting radionuclides in environmental samples [17].

Summary

New synthesis methodologies in the area of templated nanomaterials and molecular self-assembly have created an effective new type of sorbent materials that can be utilized for environmental SPME and assay application. The rigid open pore structure of SAMMS allows for the facile diffusion of the target analyte into the mesoporous matrix, thereby enabling rapid sorption kinetics. The self-assembled monolayers terminated in highly specific ligands allow the sorbent to be tailored to the specific environmental targets, providing excellent chemical selectivity and capacity. Using SAMMS SPME for selective preconcentration can enhance the selectivity and sensitivity of analytical methods. We have demonstrated XRF coupled with SAMMS for SPME to be capable of quickly determining the concentration of toxic heavy metals to levels below the EPA drinking water limits. This enhancement has also been demonstrated using electrochemical sensing [18] and radioanalytical methods. Further work developing SAMMS materials for environmental SPME and for using these new materials synergistically with analytical instrumentation is currently underway.

References

1. "Solid Phase Microextraction. Theory and Practice", J. Pawliszyn; Wiley-VCH; New York, 1997.
2. Kresge, C. T.; Leonowicz, M. E.; Roth, W. J.; Vartuli, J. C.; Beck, J. S. *Nature* 1992, *359*, 710. Beck, J. S.; Vartuli, J. C.; Roth, W. J.; Leonowicz, M. E.; Kresge, C. T.; Scmitt, K. D.; Chu, C. T. W.; Olson, D. H.; Sheppard, E. W.; McCullen, S. B.; Higgins, J. B.; Schlenker, J. L. *J. Am. Chem. Soc.* 1992, *114*, 10834.
3. "Designing Surface Chemistry in Mesoporous Silica" Fryxell, G. E., Liu, J. in Adsorption on Silica Surfaces" Papirer, E., ed., Marcel Dekker, New York, 2000.
4. Moller, K.; Bein, T. *Chem. Mater.* 1998, *10*, 2950-2963.
5. See for example: Peters, M. W.; Werner, E. J.; Scott, M. J. *Inorg. Chem.* 2002, *41*, 1707-1716. Nash, K. L.; Choppin, G. R. *Sep. Sci. Technol.* 1997, *32*, 255. Horwitz, E. P.; Kalina, D. G.; Diamond, H.; Vandergrift, G. F.;

Schulz, W. W. *Solvent Extr. Ion Exch.* **1985**, *3*, 15.   Schulz, W. W.; Horwtiz, E. P. *Sep. Sci. Technol.* **1988**, *23*, 1191.

6. Mercier, L.; Pinnavaia, T. J. *Env. Sci. Tech..* **1998**, *32*, 2749. Antonchshuk, V.; Jaroiec, M. *Chem. Mater.* **2000**, *12*, 2496-2501. Kruk, M.; Jaroniec, M.; Antochshuk, V.; Sayari, A. *J. Phys. Chem. B* **2002**, *106*, 10096-10101. Bibby, A.; Mercier, L. *Chem. Mater.* **2002**, *14*, 1591-1597. Mercier, L.; Pinnavaia, T. J. *Chem. Mater.* **2000**, *12*, 188. Mercier, L.; Pinnavaia, T. J. *Adv. Mater.* **1997**, *9*, 500.

7. "Hybrid Mesoporous Materials with Functionalized Monolayers" J. Liu, X. Feng, G.E. Fryxell, L.Q. Wang, A.Y. Kim, M. Gong, *Advanced Materials*, **1998**, *10*, 161-165. "Organic Monolayers on Ordered Mesoporous Supports" X. Feng, G.E. Fryxell, L.Q. Wang, A.Y. Kim, J. Liu, *Science,* **1997**, *276*, 923-926. Chen, X.; Feng, X.; Liu, J.; Fryxell, G. E.; Gong, M. *Sep. Sci. & Tech.* **1999**, *34*, 1121-1132.

8. K. Kemner,K. M.; Feng, X.; Liu, J.; Fryxell, G. E. ; Wang, L. -Q. ; Kim, A. Y. ; Gong, M.; Mattigod, S. V. *J. Synch. Rad.* **1999**, *6*, 633-635.

9. Mattigod, S. V.; Feng, X.; Fryxell, G. E.; Liu, J.; Gong, M. *Sep. Sci.& Tech.* **1999**, *34*, 2329-2345.

10. Fryxell, G. E.; Liu, J.; Gong, M.; Hauser, T. A.; Nie, Z.; Hallen, R. T.; Qian, M.; Ferris, K. F. *Chem. Mater.* **1999**, *11*, 2148-2154.

11. Lin, Y.; Fryxell, G. E.; Wu, H.; Englehard, M. *Env. Sci. & Tech.* **2001**, *35*, 3962-3966.

12. Birnbaum, J. C.; Busche, B.; Lin, Y.; Shaw, W. J.; Fryxell, G. E. *Chem. Commun.* **2002**, 1374-1375.

13. "Self-assembled Monolayers on Mesoporous Silica, a Super Sponge for Actinides" X. Feng, L. Rao, T. R. Mohs, J. Xu, Y. Xia, G. E. Fryxell, J. Liu, and K. N. Raymond, *Ceramic Transactions, 93, Environmental Issues and Waste Management Technologies IV,* edited by J. C. Mara and G. T. Chandler, pp. 35-42, **1999**.   Fryxell, G. E. "Actinide Specific Interfacial Chemistry of Monolayer Coated Mesoporous Ceramics" Final Report for the Environmental Managed Science Program, Project #65370, U.S. Dept. of Energy, 2001.

14. Kresge, C. T.; Leonowicz, M. E.; Roth, W. J.; Vartuli, J. C.; Beck, J. S. *Nature* **1992**, *359*, 710. Beck, J. S.; Vartuli, J. C.; Roth, W. J.; Leonowicz, M. E.; Kresge, C. T.; Scmitt, K. D.; Chu, C. T. W.; Olson, D. H.; Sheppard, E. W.; McCullen, S. B.; Higgins, J. B.; Schlenker, J. L. *J. Am. Chem. Soc.* **1992**, *114*, 10834. Zhao, D.; Huo, Q.; Feng, J.; Chmelka, B. F.; Stucky, G. D. *J. Am. Chem. Soc.* **1998**, *120*, 6024-6036. Alberius, P. C. A.; Frindell, K. L.; Hayward, R. C.; Kramer, E. J.; Stucky, G. D.; Chmelka, B. F. *Chem. Mater.* **2002**, *14*, 3284-3294. Landry, C. C.; Tolbert, S. H.; Gallis, K. W.; Monnier, A.; Stucky, G. D.; Norby, P.; Hanson, J. C. *Chem. Mater.* **2001**,

13, 1600-1608. Antonelli, D. M.; Ying, J. Y. *Curr. Op. Colloid Interface Sci.* **1996**, *69*, 131.

15. "Organic thin films and surfaces: Directions for the nineties".edited by Abraham Ulman., Academic Press, 1995.

16. Zemanian, T. S.; Fryxell, G. E.; Liu, J.; Mattigod, S.; Franz, J. A.; Nie, Z. *Langmuir* **2001**, *17*, 8172-8177. Shin, Y.; Zemanian, T. S.; Fryxell, G. E. Wang, L. Q.; Liu, J. *Microporous and Mesoporous Mater.* **2000**, *37*, 49-56.

17. "SAMMS for Solid Phase Microextraction and Assay of Radionuclides" *J. of Radioanalytical and Nuclear Chemistry,* R. Shane Addleman, Oleg B. Egorov, Matthew O'Hara, Brad Busche, Thomas S. Zemanian, Glen Fryxell, in press.

18. Yantasee, W.; Lin, Y., Zemanian, T.S., Fryxell, G.E. Voltammetric detection of lead(II) and mercury(II) using a carbon paste electrode modified with thiol self-assembled monolayer on mesoporous silica (SAMMS). *Analyst,* 128(5):467-472.   Yantasee, W.; Lin, Y.; Fryxell, G.E.; Wang, Z. "Acetamide phosphonic acid-modified carbon paste electrode: a new sensor for uranium detection" *Electroanal. Short Comm.,* (in press). Yantasee, W.; Lin, Y.; Li, X; Fryxell, G.E.; Zemanian, T.S.; Viswanathan, V. "Nanoengineered electrochemical sensor based on mesoporous silica thin-film functionalized with thiol-terminated monolayer", *Analyst,* (in press). Yantasee, W.; Lin, Y.; Fryxell, G.E.; Busche, B.J. "Simultaneous detection of cadmium(II), copper(II), and lead(II) using a carbon paste electrode modified with carbamoylphosphonic acid self-assembled monolayer on mesoporous silica (SAMMS)" *Anal. Chim. Acta,* (submitted).

Chapter 25

Photosynthetic Biosensors for Rapid Monitoring of Primary-Source Drinking Water

Elias Greenbaum[1,2,*], Miguel Rodriguez, Jr. [1],
and Charlene A. Sanders[1]

[1]Chemical Sciences Division, Oak Ridge National Laboratory, Oak
Ridge, TN 37831
[2]Center for Environmental Biotechnology and Graduate School of Genome
Science and Technology, University of Tennessee, Knoxville, TN 37996
*Corresponding author: greenbaum@ornl.gov

Introduction

Algae and cyanobacteria are present in all surface waters that are exposed to sunlight. The unattached microscopic organisms that are found individually or in small clumps floating in rivers and lakes are composed of phytoplankton and zooplankton. Most of the phytoplankton is composed of algae *(1)*. Chlorophyll *a* is the major photosynthetic pigment in algae. Since microalgae are ubiquitous and grow rapidly, quantification of this pigment can serve as a good indicator of environmental changes in aquatic ecosystems *(2)*.

The time course of fluorescence is an indication of the organism's ability to perform photosynthesis and is directly related to its physical well-being. One of the most useful parameters in fluorescence analysis from photosynthetic tissue is the efficiency of Photosystem II (PSII) photochemistry. The ratio of variable to maximal fluorescence (F_V/F_M) which is equal to the maximal fluorescence minus the minimal fluorescence divided by the maximal fluorescence (F_M-F_0)/F_M) is called the quantum yield of photochemical efficiency. It may be calculated from photosynthetic tissue that is subjected to a saturating flash of light *(3,4)*.

The purpose of this paper is to describe the application of naturally occurring photosynthetic microorganisms as biosensors for the detection of chemical antagonists in primary-source drinking water. The agents used in this study were methyl parathion (MPt), potassium cyanide (KCN), diuron (DCMU), and paraquat. Methyl parathion is an organophosphorus insecticide and a cholinesterase inhibitor that is structurally and functionally similar to the chemical warfare agents classified as nerve agents. Severe exposure in humans and animals can lead to convulsions, unconsciousness, cardiac arrest, and death. Hydrogen cyanide has an odor characteristic of bitter almonds and is completely miscible in water. For these experiments, the water-soluble salt KCN was used. The cyanide ion is an extremely toxic and fast-acting poison. Paraquat is a herbicide that is highly soluble in water. Death is usually due to progressive pulmonary fibrosis and epithelial proliferation in the lungs. DCMU (Diuron) is a substituted urea–based herbicide. Diuron is a nonionic compound with moderate water solubility. The U.S. Environmental Protection Agency has ranked Diuron fairly high (i.e., as a Priority B Chemical) with respect to potential for groundwater contamination *(5)*.

Experimental

Studies were performed using "as-is" freshwater samples and their naturally occurring populations of phytoplankton.

Cell Chamber, Fluorescence Measurements, and Toxic Agents

The experiments, illustrated in Figure 1, were designed to mimic the flow of river or lake water through the fluorescence detection system. The cell chamber/optical unit (Figure 1) is a flow system containing the fluorescence cell. Following the batch experiments, the standard fluorescence cuvette in the Walz fluorometer was replaced with the flow-through model (Hellma Cells, Inc., Model QS-131, Plainview, NY). The cuvette inlet was connected to a glass-bottle reservoir that contained the water samples, and the outlet drained to waste. This experimental arrangement allowed continuous monitoring and replacement of water samples in a manner similar to that for the contemplated operation of a real-world biosensor system. Fluorescence induction curves were measured before and during exposure to toxic agents using a Walz XE-PAM fluorometer (Walz, Effeltrich, Germany). Fluorescence excitation and emission wavelengths were 660 and 685 nm, respectively. A halogen lamp actinic light source illuminated the cuvette at an intensity of 500 $\mu E \cdot m^{-2} \cdot s^{-1}$ via a fiber-optic cable through direct connection to the cell chamber.

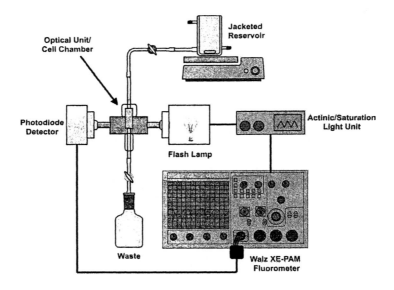

Figure 1. Experimental apparatus for continuous monitoring of water quality. Batch experiments used a static fluorescence cuvette. Continuous experiments used a flow-through cuvette.

Fluorescence induction curves were recorded every 5 min; data collection for each curve was completed within 10 s. Data extracted from the fluorescence induction curves were used to calculate F_s, F_{max}, F_v (variable fluorescence = $F_{max} - F_s$), and the efficiency of PSII photochemistry (F_v/F_{max}). A 200-mL water sample was placed in a jacketed reservoir. The sample was continuously stirred and maintained in darkness with a black cloth. The reservoir was connected to the flow-through fluorescence cell with flexible tubing. To obtain a homogenous sample before each recording, the volume in the fluorescence cell was replaced three times. After control data were collected, the volume in the reservoir was adjusted to 100 mL and the toxic agent was added. The toxic agents were prepared as stock solutions prior to addition to the reservoir and were injected directly into the top of the vessel and immediately mixed with the sample. Spent samples were drained into a waste bottle (Figure 1). Upon arrival in the laboratory from collection sites at the rivers, the water samples were kept under a fluorescent lamp at an illumination of 50 μE· m^{-2}· s^{-1} until use.

Results

Freshwater Batch Experiments

Field samples were drawn from selected locations of the Clinch River, Oak Ridge, Tennessee, and the Tennessee River, Knoxville, Tennessee. The Oak Ridge samples were taken from the Clark Center Recreation Park, the Melton Hill Hydroelectric Dam, and the Oak Ridge Marina. Table 1 summarizes data obtained from samples at various locations after 25-min exposure to the agents indicated.

Table I. Decrease in Photochemical Yields of Naturally Occurring Algae in Primary-Source Drinking Waters from the Clinch and Tennessee Rivers Following Exposure to Toxic Agents

Sample site	Toxic agent ($\Delta \pm prob.\ err.\ (\%)$) [a]		
	KCN	MPt	DCMU
Clark Center Recreation Park	22.78 ± 1.63	8.32 ± 0.21	17.71 ± 1.32
Melton Hill Hydroelectric Dam	29.85 ± 4.17	7.66 ± 0.90	23.45 ± 4.77
Oak Ridge Marina	25.88 ± 0.90	8.58 ± 0.27	12.81 ± 0.81
Tennessee River	21.89 ± 0.76	3.28 ± 0.18	14.77 ± 1.81

[a] For each agent, Δ is defined as the decrease in photochemical yield following a 25-min exposure. It was computed by subtracting the value obtained for each toxic agent from its corresponding control. The probable error is the computed error (the square root of the sum of the squares of the sample standard deviations of the toxin and control data) based on standard error analysis (6).

Continuous Experiments

Continuous experiments were performed with samples drawn from the Clark Center Recreation Park. Figure 2 presents the decrease in photochemical yield of these samples. The change in yield as a function of exposure time was normalized to the data points at 5 min. The error bars represent the standard deviations for the samples. Control and experimental runs were conducted in triplicate, except for MPt experiments, which were conducted in duplicate. All yields were computed from three runs, except for MPt yields, which were computed from data collected from two runs. Results are expressed as means ±1 S.D.

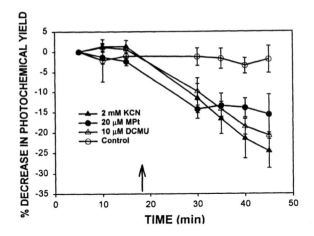

Figure 2. Decrease in photochemical yield of naturally occurring algae in Clinch River (Clark Center Recreation Park) water. The first data points were recorded after 5 min of dark adaptation. MPt, KCN, and DCMU were added at 18 min (indicated by arrow), and fluorescence data were recorded at 30, 35, 40, and 45 min.

Discussion

Our results demonstrate that naturally occurring freshwater algae can be used as biosensor material for the detection of toxic agents in sunlight-exposed primary drinking water supplies. These agents block electron transport, impair light energy transfer, or generate toxic secondary photoproducts, all of which provide signals that can trigger an alarm. Our results showed that the tissue-based biosensors experienced a decrease in photochemical yield when exposed to MPt, KCN, paraquat, and DCMU. A detectable effect was observed in every freshwater sample tested.

Conclusions

The data presented in this paper indicate that biosensors based on fluorescence induction curves of naturally occurring freshwater algae can be used to detect methyl parathion, cyanide, paraquat, and DCMU in primary water supplies under appropriate experimental conditions. In the context of current state-of-the-art biosensor research, they are unique: in the case of sunlight-exposed drinking water, the biosensors occur naturally in the medium to be protected. When combined with encrypted data telecommunication and a database-lookup library containing pertinent data for healthy algae, this approach to protection of sunlight-exposed primary drinking water supplies may be of practical value under real-world conditions.

Acknowledgements

The authors thank Drs. John W. Barton and Brian H. Davison for their assistance with the methyl parathion experiments. We also thank Dr. K. Thomas Klasson for designing spreadsheets for data analysis, Ms. Angela R. Jones and Ms. Patricia A. Wilson for secretarial support, and Dr. Barton and Ms. Marsha K. Savage for reviewing the manuscript. This research was supported by the U.S. Department of Energy and the Tissue-Based Biosensors Program, Defense Advanced Research Projects Agency, under MIPR No. 99-H250 with Oak Ridge National Laboratory. Oak Ridge National Laboratory is managed by UT-Battelle, LLC, for the U.S. Department of Energy under contract DE-AC05-00OR22725. The data contained in this paper were originally described in reference 7.

References

1. Palmer, C. M. Algae in water supplies. *Public Health Service Publication No. 657*; U.S. Department of Health, Education, and Welfare: Cincinnati, OH, 1959.
2. Pinto, A. M. F.; Von Sperling, E.; Moreira, R. M. Chlorophyll-a determination via continuous measurement of plankton fluorescence: methodology development. *Water Res.* **2001**, *35*, 3977–3981.
3. Schreiber, U.; Bilger, W.; Neubauer, C., Chlorophyll fluorescence as a nonintrusive indicator for rapid assessment of in vivo photosynthesis. *Ecol. Stud.* **1994**, *100*, 49–70.
4. Falkowski, P. G.; Raven, J. A. *Aquatic Photosynthesis*; Blackwell Science Publications: Malden, MA, 1997.
5. *Guidelines for Canadian Drinking Water Quality*, 6th ed.; Ministry of Supply and Services: Ottawa, Canada, 1996.
6. Sime, R. J. *Physical Chemistry: Methods, Techniques, and Experiments*; Saunders College Publishing: Philadelphia, PA, 1990, pp 136-1.
7. Rodriguez, M., Jr.; Sanders, C. A.; Greenbaum, E. Biosensors for rapid monitoring of primary-source drinking water using naturally occurring photosynthesis. *Biosens. Bioelectron.* **2002**, *17*, 843–849.

Environmental Applications: Treatment and Remediation Using Nanotechnology

Editors
Daniel Strongin
Wei-Xian Zhang

Chapter 26

Environmental Applications: Treatment/Remediation Using Nanotechnology: An Overview

Daniel Strongin

Department of Chemistry, Temple University, Philadelphia, PA 19122

Contributions in this chapter address the hypothesis that nanotechnology can be used for the treatment and remediation of toxins in the environment. Toward this end papers are presented that detail research on a variety of nanostructured materials ranging from natural protein structures to nanostructures produced by lithographic techniques that have already been developed in the electronics industry. The motivation behind this session was to bring to the forefront the diversity of nanostructure design and application, and to expose different approaches to researchers in a variety of scientific fields.

The chemical waste resulting from industrialized nations presents a serious set of environmental problems. The treatment and/or remediation of toxic wastes generated over many years by industry, the military, the civilian sector, and national laboratories is a daunting challenge. It is not surprising that an intense scientific effort has gone into the development of chemical, biochemical, and photochemical remediation schemes to remove or destroy toxins in soils and waters. Common degradation techniques, for example, implemented to remediate waste waters include zero valent iron chemistry*(1)*, photodetoxification using UV/Ozone systems *(2)*, biodegradation *(3,4)*, and phytoremediation *(5,6)*.

A working hypothesis by researchers in the nano-arena has been that nano-materials may provide a chemistry conducive to environmental remediation that cannot be obtained at more traditional spatial dimensions (i.e., $> \mu m$). The *Environmental Applications: Treatment/Remediation using Nanotechnology* symposium was an ambitious attempt to bring many different nanotechnology based remediation and treatment strategies together into one session. The session not only highlighted the activity in this area, but emphasized the unique structural properties of nanoparticles that set them apart from more traditional size particles.

The first session was started by Dr. Jose Rodriguez of Brookhaven National Laboratory who gave an overview of the application of nanogold particles supported on TiO_2 for SO_2 destruction. TiO_2 is a commonly used catalyst in industry for SO_2 removal, but it is shown in the first talk that nanogold particles has the potential to improve the performance of TiO_2 alone. The nanogold-TiO_2 system shows a very high efficiency toward S-O bond cleavage, a necessity in SO_2 destruction. The reactivity of this system stems from the unique electronic structure of the nanogold supported on TiO2, emphasizing the new realm of chemical reactivity available at the nanoscale. Not all metals, however, are easily deposited on solid supports as well defined nanoparticles so that new synthetic strategies are being pursued by researchers. Professor Somorjai of the University of California followed Dr. Rodriguez's talk by giving an overview of using modern lithography to prepare patterned metal nanostructured on solid supports. Specifically, Professor Somorjai uses electron beam lithography, to produce platinum nanoparticles between 5 and 50 nm. Professor Somorjai also emphasized that nanotechnology is intimately intertwined with catalysis, since industrial catalysts are often of nanosize and nature's catalysts, enzymes, are composed of inorganic clusters surrounded by high molecular weight protein. Professor Somorjai argued that advances in nanotechnology could help achieve an sought after goal in the area of catalysis; the achievement of close to 100% catalyst selectivity often found in natural enzymatic systems. It would be advantageous apply the high selectivity of enzymes to remediation efforts. Due to the expense and poor stability of enzymes in remediation efforts, their use is problematic. Dr. Jungbae Kim of Pacific Northwest National Laboratory gave a exciting talk that gave an overview of stabilizing enzymatic activity by using enzyme-polymer composites in nano-meter scale or single-enzyme scale. These researchers crosslinked a composite silicate shell around the individual enzyme molecules yielding a highly stabilized system. Professor Martin Schoonen followed this talk with a study that characterized the charge development of the biologically relevant protein ferritin in environmentally relevant solutions. Ferritin, which has a roughly spherical shape, has an inorganic core composed of nanosized iron oxyhydroxide. In a followup talk, Hazel-Ann Hosein of Temple University concluded the first session by showing that ferritin could be used as a precursor for the formation of nano iron metal and oxyhydroxide. In essence, ferritin could be used to template the growth of potentially useful nanoparticles.

References

1. Xu, Y.; Zhang, W.-X. *Hazardous and Industrial Wastes* **1999**, *31st*, 231-239.
2. Hoffmann, M. R.; Martin, S. T.; Choi, W.; Bahneman, D. W. *Chemical Reviews* **1995**, *95*, 69-96.
3. Karamanev, D. G. *J. Sci. Ind. Res.* **1999**, *58*, 764-772.
4. Starrett, S.; Bhandari, A.; Xia, K. *Water Environ. Res.* **1999**, *71*, 853-860.
5. Teaca, C.-A. *Cellul. Chem. Technol.* **1999**, *33*, 351-352.
6. Gadd, G. M. *J. Chem. Technol. Biotechnol.* **2001**, *76*, 325.

Chapter 27

Activation of Gold Nanoparticles on Titania: A Novel DeSOx Catalyst

José A. Rodriguez

Department of Chemistry, Brookhaven National Laboratory, Upton, NY 11973 (rodriguez@bnl.gov)

Abstract

The DeSO$_x$ activity of Au nanoparticles supported on titania and magnesium oxide is examined and compared. The Au/TiO$_2$ system exhibits a remarkable activity for the destruction of SO$_2$. Extensive dissociation of SO$_2$ is observed on Au/TiO$_2$(110), with negligible dissociation on Au/MgO(100). Similar trends are found in Au/TiO$_2$ and Au/MgO high surface area catalysts. These results illustrate the importance of the oxide support for the activation of gold nanoparticles.

The destruction of SO_2 ($DeSO_x$) is a very important problem in environmental chemistry *(1,2)*. SO_2 is frequently formed during the combustion of fossil-derived fuels in factories, power plants, houses, and automobiles *(1)*. Every year the negative effects of acid rain (main product of the oxidation of SO_2 in the atmosphere) on the ecology and corrosion of monuments or buildings are tremendous. Thus, new environmental regulations emphasize the need for more efficient technologies to destroy the SO_2 formed in combustion processes *(2)*. Titania is the most common catalyst used in the chemical industry and oil refineries for the removal of SO_2 through the Claus reaction: $SO_2 + 2H_2S \rightarrow 2H_2O + 3S_{solid}$ *(2)*. Different approaches are being tested for improving the performance of titania in DeSOx operations. We have found that the addition of gold to TiO_2 produces desulfurization catalysts with a high efficiency for the cleavage of S-O bonds. In this respect, the Au/TiO_2 system is much more chemically active than either pure titania or gold *(3)*.

Bulk metallic gold typically exhibits a very low chemical and catalytic activity *(4)*. Among the transition metals, gold is by far the least reactive and is often referred as the "coinage metal". Recently, gold has become the subject of a lot of attention due to its unusual catalytic properties when dispersed on some oxide supports (TiO_2, CrO_x, MnO_x, Fe_2O_3, Al_2O_3, MgO) *(3,5-9)*. Several models have been proposed for explaining the activation of supported gold: from special electronic properties resulting from the limited size of the active gold particles (usually less than 10 nm), to the effects of metal-support interactions (i.e. charge transfer between the oxide and gold). In principle, the active sites for the catalytic reactions could be located only on the supported Au particles or on the perimeter of the gold-oxide interface. To address some of these issues, the reactivity of $Au/TiO_2(110)$ and $Au/MgO(100)$ towards SO_2 was compared *(3,10)*.

Experimental Methods

Photoemission was used to study the chemistry of SO_2 on the supported nanoparticles. The experiments were performed in a standard ultra-high vacuum chamber (base pressure ~ 6×10^{-10} Torr) that is part of the U7A beamline of the National Synchrotron Light Source (NSLS) at Brookhaven National Laboratory *(3)*. The overall instrumental resolution in the photoemission experiments was approximately 0.35 eV. The $Au/TiO_2(110)$ and $Au/MgO(100)$ surfaces were prepared as described in refs 3 and 10. SO_2 was dosed through a glass-capillary array doser positioned to face the sample at a distance of ~ 2 mm. The SO_2 exposures are based on the ion gauge reading and were not corrected for the capillary array enhancement (> 5 enhancement factor with respect to background dosing).

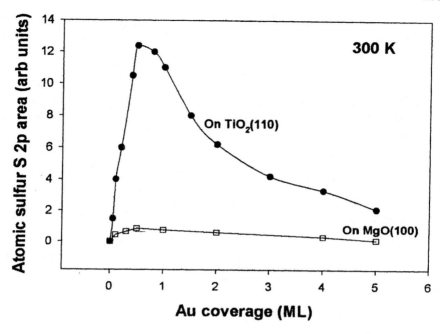

Fig 1. Amount of atomic sulfur generated after dosing SO$_2$ to a series of Au/TiO$_2$(110) and Au/MgO(100) surfaces at 300 K. The coverage of sulfur is assumed to be proportional to the area under the S 2p features in photoemission.

RESULTS

Figure 1 compares S 2p areas measured for atomic S after dosing SO$_2$ (5 langmuir) to Au/MgO(100) and Au/TiO$_2$(110) surfaces at 300 K. Neither MgO(100) nor TiO$_2$(110) are able to dissociate SO$_2$ on their own. On both oxide supports, the largest activity for the full dissociation of SO$_2$ is found in systems that contain Au coverages smaller than 1 ML when the size of the Au nanoparticles is below 5 nm *(3,6,10)*. Clearly the Au/TiO$_2$(110) system is much more chemically active than the Au/MgO(100) system. These data indicate that titania either plays a direct active role in the dissociation of SO$_2$ or modifies the chemical properties of the supported Au nanoparticles *(3,10)*.

Photoemission results and TPD data indicate that Au particles supported on MgO(100) bond SO_2 substantially stronger than extended surfaces of gold *(10)*. The heat of adsorption of the molecule on the Au nanoparticles is ~ 15 kcal/mol compared to 8 kcal/mol for SO_2 on Au(111). Density functional calculations indicate that the enhancement in the SO_2 adsorption energy is simply due to the presence of corner sites (i.e. Au atoms with a low coordination number) in the nanoparticles (10). The dissociation of SO_2 on Au/MgO(100) is very limited due to weak Au/MgO(100) interactions.

Recent catalytic tests for the Claus reaction (SO_2 + $2H_2S$ → $2H_2O$ + $3S_{solid}$) and the reduction of SO_2 by carbon monoxide (SO_2 + $2CO$ → $2CO_2$ + S_{solid}) show that the Au/TiO_2 system is 5-10 times more active than pure TiO_2 *(11)*. Interestingly, polycrystalline foils of gold are inactive as catalysts for these DeSOx processes. In the case of Au/MgO, a limited DeSO$_x$ activity is observed *(11)*.

Acknowledgements

The studies described here were done in collaboration with Z. Chang, J. Dvorak, J. Evans, L. Gonzalez, J. Hrbek, T. Jirsak, G. Liu, A. Maiti, and M. Perez. This research was carried out at Brookhaven National Laboratory and supported by the US Department of Energy.

References

1. Slack, A.V.; Holliden, G.A. Sulfur Dioxide Removal from Waste Gases, 2nd ed ; Noyes Data Corporation: Park Ridge, NJ 1975.
2. Pieplu, A.; Saur, O.; Lavalley, J.-C.; Legendre, O.; Nedez, C. Catal. Rev.-Sci. Eng. 1998, 40, 409.
3. Rodriguez, J.A.; Liu, G.; Jirsak, T.; Hrbek, J.; Chang, Z.; Dvorak, J.; Maiti, A. J. Am. Chem. Soc. 2002, 124, 5242.
4a. Thomas, J. M.; Thomas, W. J. Principles and Practice of Heterogeneous Catalysis; VCH: New York, 1997
b. Somorjai, G.A. Introduction to Surface Chemistry and Catalysis; Wiley: New York, 1994.
5. Haruta, M. Catal. Today, 1997, 36, 153.
6. Valden, M.; Lai, X.; Goodman, D.W. Science, 1998, 281, 1647.
7. Rodriguez, J.A.; Chaturvedi, S.; Kuhn, M.; van Ek, J.; Diebold, U.; Robbert, P.S.; Geisler, H.; Ventrice, C.A. J. Chem. Phys. 1997, 107, 9146.

8.   Bondzie, V.A.; Parker, S.C.; Campbell, C.T. J. Vac. Sci. Technol. A, 1999, 17, 1717

9.   Campbell, C.T. Curr. Opin. Solid State Mater. Sci. 1998, 3, 439.

10.   Rodriguez, J.A.; Perez, M.; Jirsak, T.; Evans, J.; Hrbek, J.; Gonzalez, L. Chem. Phys. Lett., 2003, in press.

11.   Evans, J.; Lee, H; Fischer, I. research in progress.

Chapter 28

Fabrication of Two-Dimensional and Three-Dimensional Platinum Nanoparticle Systems for Chemisorption and Catalytic Reaction Studies

Gabor A. Somorjai

Department of Chemistry and Lawrence Berkeley National Laboratory, University of California, Berkeley, CA 94720

The aim of catalysis science in the 21st Century is the development of catalyst systems that exhibit 100% selectivity toward the desired product. Two-dimensional model catalysts are fabricated by electron beam lithography and size-reduction photolithography with precise control of metal catalyst nanoparticle size, surface structure, location and oxide-metal interface. 3-dimensional platinum nanoparticle catalyst systems are being developed as well to achieve high surface area and precise control of all of the structural parameters.

Catalysis: The Central Field of Nanoscience and Nanotechnology

Background

There are two types of catalysts that carry out chemical reactions with high rates and selectivity *(1)*. Enzymes are nature's catalysts, and many of them are composed of inorganic nanoclusters surrounded by high molecular weight proteins. These catalysts help the human body to function and are responsible for the growth of plants. They usually operate at room temperature and in aqueous solution. Synthetic catalysts, either heterogeneous or homogeneous, are often metal nanoclusters that are used in the chemical technologies to carry out reactions with high turnover and selectivity. The heterogeneous systems have high catalyst surface area and they frequently operate at high temperatures in the 400-800K range. The reactants and products flow by the catalyst interface in the gas phase, if possible, to facilitate high reaction rates and removal of the product molecules from the catalyst bed. Both enzyme and synthetic catalysts are usually nanoclusters in size and thus the fields of catalysis science and technologies are also nanoscience and nanotechnologies. The evolution of the field of catalysis is strongly coupled to the development of nanoscience and nanotechnology at the present.

Catalysis in the 20th Century focused primarily on activity, increasing turnover rates to produce more molecules/unit area/unit time *(2)*. Selectivity was of lesser concern because the disposal of undesirable byproducts as waste was not costly. This has all changed in the 21st Century, because waste disposal is now expensive. As a result, the focus of catalysis science at present is to achieve 100% selectivity in all catalyst-based chemical processes. This is often called green chemistry.

The Known Molecular Ingredients of Catalytic Activity and Selectivity

Results

There are six important identifiable features that influence both catalyst activity and selectivity. These are 1) metal surface structure; 2) bonding modifier additives; 3) mobility of the metal clusters to restructure as well as the mobility of adsorbates on these clusters; 4) selective site blocking; 5) bifuntional catalysis and 6) oxide-metal interface sites. Examples were given of how these catalyst ingredients work to produce improved reactivity.

Fabrication of High-Technology Catalysts

In order to obtain high selectivity toward the ultimate goal of 100% selectivity, which is a road map for obtaining selective and green catalysts for chemical processes, one has to assert molecular control over the size, location, structure, and promoters of the catalysts. We are attempting to do this by fabricating 2-dimensional and 3-dimensional catalysts.

2-Dimensional and 3-Dimensional Catalyst Fabrication

Using electron beam lithography (5), one can obtain nanoparticles of platinum in the size range of 5-50 nm. An atomic force microscope picture of such a structure is shown in Figure 1. Using electron microscopy, both in the back reflection and in the transmission modes, one can image the nanoparticle array and the structure of each nanoparticle. It was found that upon heating these nanoparticles either in vacuum or in hydrogen or oxygen, they restructured to form highly-perfect single crystals. Methods like this allow us to prepare 10^9 nanoparticles in about a day. This amounts to a surface area of 1 mm^2, which is quite small for studies of reaction selectivity. We need about 10^{11} nanoparticles, which would provide a cm^2 surface area, which is adequate to study, for example, reforming reactions or CO hydrogenation reactions. However, that would take 100 days to produce by electron beam lithography. As a result, we have turned to photolithography. A new process called size reduction lithography (6) has been developed in Berkeley permitting us, by selective etching, to reduce the size from 600 nm photolithographically-prepared features down to 7 nm size features. The scheme of this process is shown in Figures 2 and 3. Figure 2 shows the overall scheme, while in Figure 3, the steps that lead to a 50% reduction in size through the first cycle of size reduction lithography is shown. Repetition of such a cycle several times permits us to reach 7 nm in size. Figure 4 and 5 show electron microscopy pictures of the nanowires produced this way, before and after the first size reduction cycle. Once such a nanowire or nanodot mold has been produced, we use polymer imprinting technology (nanoimprint lithography) to press the silicon nanowires or nanodots into PMMA at elevated temperatures under pressure to produce an image, which upon plasma etching and metal deposition, leads to the formation of transition metal catalyst nanowires or nanodots. This way we can prepare, on a 4″ silicon wafer, a 1 cm^2 surface area nanocatalyst assembly, which amounts to 10^{11} nanowires or nanodots.

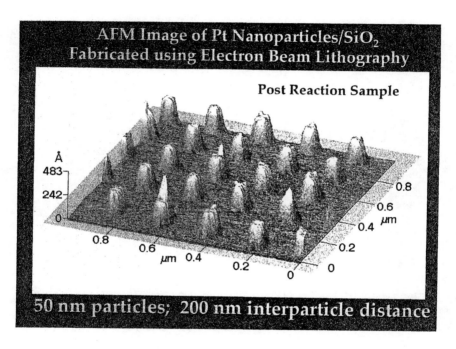

Figure 1: AFM image of 50 nm Pt nanoparticles with 200 nm spacing on SiO₂, fabricated by electron beam lithography.

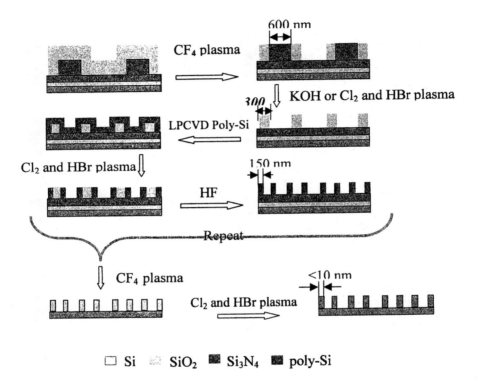

600 nm

CF₄ plasma

KOH or Cl₂ and HBr plasma

300

LPCVD Poly-Si

Cl₂ and HBr plasma

150 nm

HF

Repeat

CF₄ plasma

Cl₂ and HBr plasma

<10 nm

☐ Si ▨ SiO₂ ▇ Si₃N₄ ▇ poly-Si

Figure 2: Scheme for multiple size reduction lithography.

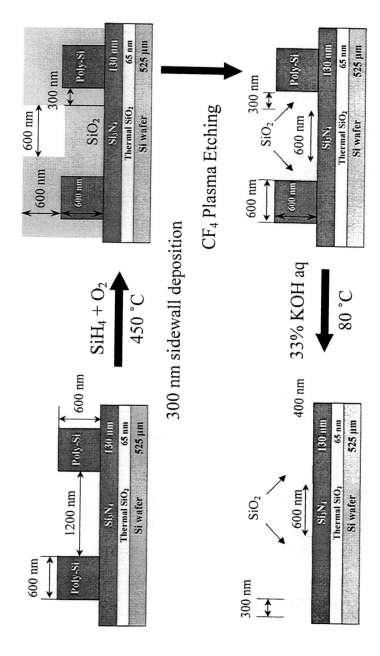

Figure 3. Focus on the steps of size reduction lithography that results in a 50% reduction in feature size.

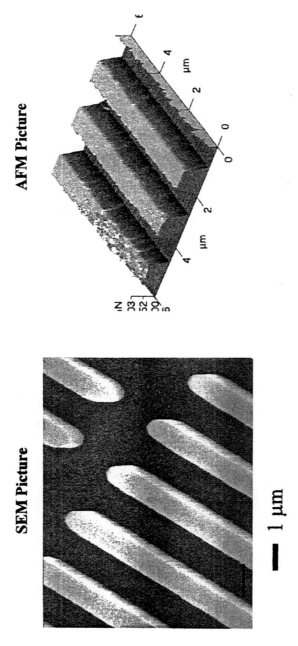

AFM Picture

SEM Picture

1 µm

Figure 4: SEM and AFM topography image of photolithographically-defined poly-Si pattern.

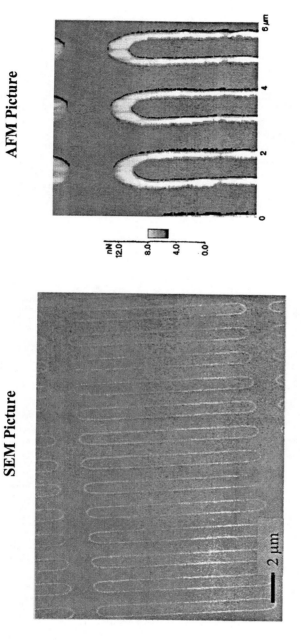

Figure 5: SEM and AFM topography image of SiO₂ nanowire pattern after 1st size reduction lithography process.

218

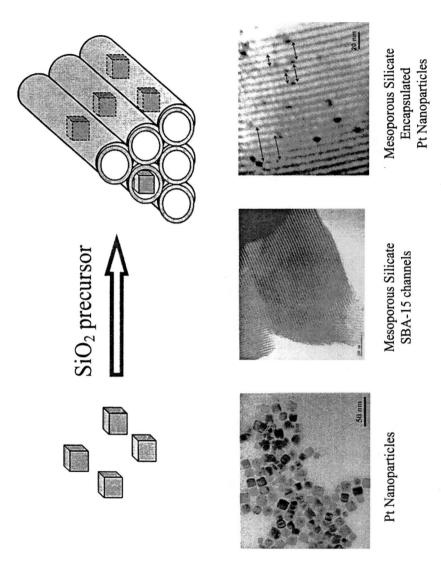

SiO₂ precursor

Pt Nanoparticles

Mesoporous Silicate
SBA-15 channels

Mesoporous Silicate
Encapsulated
Pt Nanoparticles

Figure 6: TEM image of Pt nanoparticles encapsulated in mesoporous silicate.

3-Dimensional Catalyst Fabrication

If we are to aim for higher catalyst surface area for technological applications of 1 m^2 or higher, it will require the production of 10^{15} nanoparticles or greater in number. This can only be obtained in 3-dimensional structures. Figure 6 shows our approach *(7)*. Using a sol-gel technique, we produce nanoparticles of equal size and structure that are capped by polymers. This fabrication is carried out in solution. In the same solution, we then place precursor molecules to form silicates, which, under proper conditions, produce mesoporous silicates that encapsulate the nanoparticles in the mesopores as they form. This way, we can place large concentrations of transition metal nanoparticles inside mesoporous solids to achieve high loading as well as precise size and surface structure control. These approaches are aimed toward producing nanoparticles for very high selectivity catalytic reactions.

It should be noted that enzymes exhibit 100% selectivity in important catalytic reactions, and these enzymes operate at room temperature and in aqueous solutions. Future studies will correlate the chemistry and mechanism of enzyme catalysis with those of synthetic catalysts under the same conditions at room temperature and in aqueous solutions. Such correlated studies are promising to teach us how nature prepares highly selective nanocluster catalysts, and therefore how to synthesize and use them in chemical technology.

Acknowledgement

This work was supported by the Director, Office of Science, Office of Basic Energy Sciences, Chemical Sciences Division, of the U.S. Department of Energy under Contract No. DE-AC03-76SF00098.

References

1. Somorjai, G.A., *Catal. Lett.* 76 (2001) 111-124.
2. Somorjai, G.A.; McCrea, K., *Appl. Catal. A-Gen.* 222 (2001) 3-18.
3. Borodko, Y.; Somorjai, G.A., *Appl. Catal. A-Gen.* 186 (1999) 355-362.
4. Ramos, A.L.D.; Kim, S.H.; Chen, P.L.; Song, J.H.; Somorjai, G.A., *Catal. Lett.* 66 (2000) 5-11.
5. Grunes, J.; Zhu, J.; Anderson, E.A.; Somorjai, G.A., *J. Phys. Chem. B* 106 (2002) 11463-11468.
6. Choi, Y.K.; Zhu, J.; Grunes, J.; Bokor, J.; Somorjai, G.A., *J. Phys. Chem. B* 107 (2003) 3340-3343.
7. Konya, Z.; Puntes, V.F.; Kiricsi, I.; Zhu, J.; Alivisatos, P.; Somorjai, G.A., *Catal. Lett.* 81 (2002) 137-140.

Chapter 29

Nano-Biotechnology in Using Enzymes for Environmental Remediation: Single-Enzyme Nanoparticles

Jungbae Kim[1,2] and Jay W. Grate[1,3]

[1]Pacific Northwest National Laboratory, 902 Battelle Boulevard, P.O. Box 999, Richland, WA 99352
[2]Jungbae.Kim@pnl.gov
[3]jwgrate@pnl.gov

We have developed armored single-enzyme nanoparticles (SENs), which dramatically stabilize a protease (α-chymotrypsin, CT) by surrounding each enzyme molecule with a porous composite organic/inorganic shell of less than a few nanometers thick. The armored enzymes show no decrease in CT activity at 30°C for a day while free CT activity is rapidly reduced by orders of magnitude. The armored shell around CT is sufficiently thin and porous that it does not place any serious mass-transfer limitation of substrate. This unique approach will have a great impact in using enzymes in various fields, including environmental remediation.

There is a growing interest in using enzymes for remediation purposes (1-2). Recent biotechnological advances in enzyme isolation and purification procedures together with recombinant technology have allowed the production of enzymes at less cost (2). The advantages of enzymatic remediation as compared to conventional microbial remediation can be summarized as follows (1)

- More harsh operational conditions (contaminant concentration, pH, temperature, and salinity)

220   © 2005 American Chemical Society

- Application to recalcitrant compounds
- No requirement of nutrients
- No requirement of biomass acclimation
- No formation of metabolic by-products
- Much reduced mass-transfer limitation on contaminants compared to microorganisms
- Easy-to-control process
- Effective in small quantity

The organic compounds, generally targeted for enzymatic remediation, are phenols, polyaromatics, dyes, chlorinated compounds, organophosphorous pesticides or nerve agents, and explosives (TNT, trinitrotoluene). Various enzymes can be used for remediation, and representative candidates are peroxidases, laccases, tyrosinases, dehalogenases, and organophosphorous hydrolases. Even though there is more and more recognition of the usefulness of enzymatic remediation, the cost for enzymes is still expensive, especially, due to the poor stability and short lifetime of enzymes in general. Enzyme stabilization can make enzymatic remediation cost-effective by increasing the lifetime of enzymes.

In this paper, we report the development of enzyme-polymer composites in nano-meter scale or single-enzyme scale, resulting in a stabilization of enzyme activity. This new approach is distinct from immobilizing enzymes on preformed solids or entrapping them in sol-gels, polymers, or bulk composite structures. Instead, the process begins from the surface of the enzyme molecule, with covalent reactions to anchor, grow, and crosslink a composite silicate shell around each separate enzyme molecule (Fig. 1). The detailed synthetic approach is as follows. A vinyl-group functionality is grafted onto the enzyme surface by covalently modifying the amino groups on the enzyme surface with acryloyl chloride. These modified enzymes are solubilized in an organic solvent such as hexane, as reported previously by Dordick et al. (3-7). The solubilized enzymes are mixed with silane monomers containing both vinyl groups and trimethoxysilane groups. Under suitable conditions, free-radical initiated vinyl polymerization yields linear polymers that are covalently bound to the enzyme surface. Careful hydrolysis of the pendant trimethoxysilane groups followed by condensation of the resulting silanols yields a crosslinked composite silicate coating around each separate enzyme molecule. We call these armored, single-enzyme nanoparticles (SENs).

We have synthesized SENs containing CT (SEN-CT) using methacryloxypropyltrimethoxysilane (MAPS) as the vinyl monomer. The products were extracted into cold aqueous buffer solution, aged at 4°C overnight for silanol condensation, filtered, and washed. The wet residue was resuspended in buffer and stored in the refrigerator. The yields of enzyme activity in the form of SENs were 38-73%.

High resolution transmission electron microscopy (TEM) images confirmed the presence of individual enzyme particles with a composite shell (Fig. 2). The seemingly hollow center of the nanoparticle, which results from the transparency of the core protein to the electron beam, matches the size and shape of the CT (8). On the other hand, the dark image surrounding the CT is the armored shell, and the presence of silicon was confirmed by energy dispersive x-ray analysis in the TEM instrument. The size of a few particles is bigger than that of single enzyme molecule, due to the co-immobilization of a few enzyme molecules in a single particle.

The CT enzymes armored as SENs exhibited impressive catalytic stability compared to free CT. They did not show any decrease in CT activity in a buffer solution at 30°C for a day while free CT is inactivated very rapidly by autolysis under the same conditions (Fig. 3). The storage stability of the SEN-CT in buffer solution was also impressive. After three-months of storage in the refrigerator (4°C), 82% of the SEN-CT activity remained in solution as determined by measurements on solution aliquots. However, it was noted that a transparent layer of SEN-CT was built up on the inner surface of the glass vial. This immobilized layer exhibited considerable enzyme activity after removal of the solution. Thus, the apparent decrease in solution enzyme activity may be accounted for, at least in part, by immobilization on the vial surface rather than by enzyme degradation.

The activities of SEN-CT and free CT were measured by the hydrolysis of N-Succinyl-Ala-Ala-Pro-Phe p-nitroanilide (TP) in a buffer solution (10 mM phosphate buffer, pH 7.8). Kinetic constants (k_{cat}, K_m, and k_{cat}/K_m) obtained via nonlinear regression based on the least squares method are given in Table 1. The catalytic efficiency (k_{cat}/K_m) of CT in SEN form was decreased by about half, and it was due to reduced k_{cat}. However, the apparent binding constant (K_m) of SEN-CT was almost the same as that of free CT. This suggests that the armored shell did not cause any significant mass-transfer limitation for the substrate (TP).

In summary, we have developed a new form of stabilized enzyme using CT as the initial example. Dramatic activity stabilization was achieved with minimal substrate mass-transfer limitation. The development of stabilized enzymes as soluble individual enzyme particles provides the opportunity to further process these new nanomaterials, in contrast to enzymes entrapped in bulk solids. The SENs can potentially be immobilized on solid supports, assembled with other nanoparticles or molecules as part of multifunctional nanomaterials, and their surfaces can be modified by silane reagents. The SEN provides a stabilized enzyme form that can penetrate and be immobilized within nanostructured or nanoporous substrates. These new hybrid enzyme nanostructures offer great

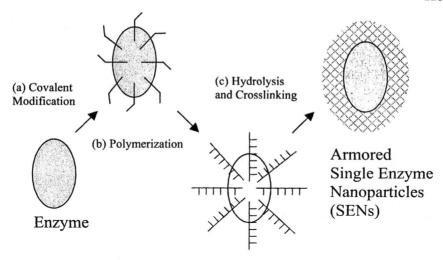

Fig. 1. *Schematic for SEN synthesis. The SEN synthesis proceeded via a three-step process: the first step is to modify the enzyme surface and solubilize the modified enzyme into a hydrophobic solvent (such as hexane); the second step is vinyl-group polymerization in hexane, forming linear polymers grafted to the enzyme surface; and, the final step is the hydrolysis and condensation of pendant alkoxysilanes, crosslinking the grafted polymers and forming the armored shell around the enzyme molecule.*

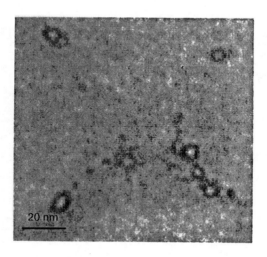

Fig. 2. *TEM image of SENs containing CT (SEN-CT). The sample was extensively washed by distilled water on a membrane filter (MWCO 10K) for the best contrast between the hollow image of CT and the dark image of armored shell. The hollow images of most particles have a dimension of CT molecule (4 x 3.5 x 8 nm) (8). The scale bar represents 20 nm.*

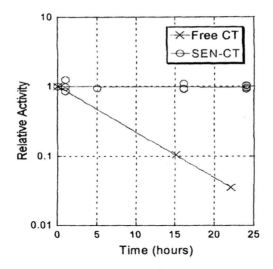

Fig. 3. Stability of SEN-CT (O) and free CT (✗). Residual activity was determined by the hydrolysis of TP in aqueous buffer (10 mM sodium phosphate, pH 7.8) after incubation at 30°C, and the relative activity is defined by the ratio of residual activity at each time point to the initial activity of each sample. All the incubations were done in plastic tubes since SEN-CT can become immobilized on the inner surface of glass vials.

Table 1. Activity of Free CT and SEN-CT.[a]

Sample	k_{cat} (s^{-1})	K_m (μM)	k_{cat}/K_m (x 10^5 $M^{-1}s^{-1}$)
Free CT	30	39	7.7
SEN-CT	14	40	3.4
SEN-CT/Free CT	0.47	1.0	0.44

[a] The CT activity was determined by the hydrolysis of TP (1.6–160 μM) in an aqueous buffer (10 mM phosphate, pH 7.8) at room temperature (22°C). Kinetic constants were obtained by using software (Enzyme Kinetics Pro from ChemSW, Farifield, CA) that performs nonlinear regression based on the least square method. The active site concentrations were determined by the MUTMAC assay.

potential to be used as prepared or processed into other forms for various applications such as bioremediation, biosensors, detergents, and enzymatic synthesis in pharmaceutical and food industries.

Acknowledgements

We thank A. Dohnalkova and C. Wang for the electron microscopy. Our work on single enzyme nanoparticles has been funded by Battelle Memorial Institute, U.S. Department of Energy (DOE) LDRD funds administered by the Pacific Northwest National Laboratory, and the DOE Office of Biological and Environmental Research under the Environmental Management Science Program. The Pacific Northwest National Laboratory is a multiprogram national laboratory operated for the U.S. DOE by Battelle Memorial Institute.

References

1. N. Durán, E. Esposito, *Appl. Catal., B* **28**, 83 (2000).
2. J. Karam, J. A. Nicell, *J. Chem. Technol. Biotechnol.* **69**, 242 (1997).
3. P. Wang, M. V. Sergeeva, L. Lim, J. S. Dordick, *Nature Biotechnol.* **15**, 789 (1997).
4. J. S. Dordick, S. J. Novick, M. Sergeeva, *Chem. Ind.* **1**, 17 (January 1998).
5. S. J. Novick, J. S. Dordick, *Chem. Mater.* **10**, 955 (1998).
6. J. Kim, R. Delio, J. S. Dordick, *Biotechnol. Prog.* **18**, 551 (2002).
7. V. M. Paradkar, J. S. Dordick, *J. Am. Chem. Soc.* **116**, 5009 (1994).
8. H. Tsukada, D. M. Blow, *J. Mol. Biol.* **184**, 703 (1985).

Chapter 30

Charge Development on Ferritin: An Electrokinetic Study of a Protein Containing a Ferrihydrite Nanoparticle

Mark Allen[1], Trevor Douglas[1], Danielle Nest[2], Martin Schoonen[2], and Daniel Strongin[3]

[1]Department of Chemistry and Biochemistry, Montana State University, Bozeman, MT 59717
[2]Department of Geosciences, Stony Brook University, Stony Brook, NY 11794
[3]Department of Chemistry, Temple University, Philadelphia, PA 19122

Introduction

The iron storage protein ferritin is ubiquitous in biological systems where it functions to sequester iron as a nanoparticle of ferric oxyhydroxide, encapsulated within a protein cage-like structure. Ferritin has several interesting physical characteristics that may lead to new remediation techniques in the future. Both the natural ferritin, synthetic ferritin-like materials, as well as ferritin-derived nano particles have potential use in new remediation techniques. For example, ferritin has already been shown to promote the reduction of hexavalent chromium dissolved in aqueous solutions *(1)*. In addition, by changing the functionality of the surface groups, the surface charge and hydrophobicity of cage can be controlled *(2)*. Hence, it may be possible to tailor the properties of the protein cage for a particular remediation application. Finally, apo-ferritin (i.e., ferritin without a mineral nano particle) can be used for the synthesis of nanometer-sized metal oxyhydroxide or metal sulfide particles *(3-6)*. The protein cage can be removed after synthesizing the nm-size particle within the cage.

In this contribution, we report our first results of a collaborative study of the surface charge development of ferritin as a function of solution conditions. This represents a small, but important component of our research program. Similar to contaminant-colloid interactions *(7)*, interaction of dissolved ionic

226

contaminants with ferritin may be influenced or dictated by the net surface charge of the protein. In addition, binding of ionic species, for example metal ions, can change the surface charge of the protein. Hence, in designing remediation strategies based on ferritin or ferritin-like materials, it is important to understand their surface charge development as function of solution composition.

Ferritin

The ferritin protein cage is comprised of 24-subunits, which assemble into a cage-like architecture roughly 12 nm in diameter *(8,9)*. In the assembled structure, the protein subunits of ferritin present charged amino acid side chains on the outer surface. These acidic (glutamic acid, aspartic acid) and basic residues (arginine, lysine, histidine) dominate the charge characteristics of the protein. Calculations of the electrostatic potential of the assembled protein reveal delocalized charge over the surface of the protein with some localized regions of high charge density *(10)*. The overall surface charge of the protein is highly pH dependent due to acid-base ionization of the surface exposed residues. The pH at which the horse-spleen ferritin protein carries no net charge, the isoelectric point (pI), has been reported as 4.5-4.7, when measured by isoelectric focusing *(11)*. Here, we report the measurement of the isoelectric point of native ferritin (containing Fe_2O_3) by measuring the zeta potential as a function of solution pH.

Light scattering-based Zetapotential Measurements

The effective surface charge, at the shear plane of a colloidal particle, can be measured directly as the zeta potential. This is accomplished by measuring the electrophoretic mobility of the particle (μ), during application of an electric field. Given the size of the ferritin, 12 nm diameter, dynamic light scattering techniques must be used to determine the electrophoretic mobility *(12,13)*. There are two light scattering-based techniques to measure electrophoretic mobilities. These are Laser Doppler Velocimetry (LDV) and Phase Analysis Light Scattering (PALS). LDV is routinely used and electrophoretic mobility is related to changes in the correlation function as a result of an applied electrical field. The method requires the use of a DC field on the other of several volts per cm and is limited to solutions with ion strengths below about 10 mmol/L. By contrast, PALS-based zetapotential measurements can be obtained on suspensions with ion strengths as high as 3 M KCl. In this study we report zetapotential data obtained using the PALS method.

Experimental

Native horse spleen ferritin (Calzyme, SanLuis Obispo CA) containing a ferrihydrite ($Fe_2O_3 \cdot nH_2O$) mineral core was purified by size exclusion chromatography (Superose 6, Pharmacia-Amersham) and monitored both at 280 nm (protein) and 415 nm (Fe-oxide). Prior to investigation of the zeta potential, the purified protein was dialyzed into an appropriate buffer (50 mM acetate: pH 4.0, 4.5, 5.0, MES: pH 5.5, 6.0, 6.5) or into a salt background (1 mM KCl). The pH of the low salt sample was adjusted by first concentrating the sample in a Microcon ultrafiltration unit and diluting back to the original volume with the appropriate buffer (1 mM) and checking the pH.

The electrophoresis measurements were conducted in two different laboratories (Montana State University and Stony Brook University) on the exactly the same ferritin with identical equipment. The equipment in both laboratories is a Brookhaven Instruments Corporation ZetaPlus equipped with a red diode laser (671 nm) and PALS option.

Results

The AC zeta potential measurements were performed on mineralized HS ferritin samples at both low (1mM) and high (50 mM) ionic strength. Prior to zeta potential measurements, the protein was first purified by size exclusion chromatography and analysed by dynamic light scattering and size exclusion chromatography to confirm that the sample was monodisperse, with a diameter of 12 nm. As shown in Figure 1, the change in the measured zeta potential as a function of pH follows the typical sigmoidal shape of a titration curve in the pH range from 3 to 7. The measured data could be fit to a sigmoidal curve. There was no significant difference between the curves measured by investigators at the two institutions. In addition, there was no significant difference between measurements made at low and high ionic strength. At a pH of 4.3 to 4.4, the ferritin samples were found to have no net charge, which corresponds to the isoelectric point (pI) of these protein assemblies. This was independent of the ionic strength as shown in Figure 1 measured, at 1 mM and 50 mM respectively, using the AC measurement. This is only slightly lower than the pI value reported from isoelectric focusing methods (11).

The results reported here are only the first step toward understanding the interaction of dissolved metal ions and ferritin. Using the results presented here as a reference, we are now conducting electrokinetic experiments to evaluate the interaction of dissolved metal ions, including chromate, with ferritin. The notion is that if a metal ion forms a chemical bond with the sorbent (here ferritin), it will change the sorbent's surface charge. We have used this research strategy before to study the interaction of aqueous metal ions and aqueous organic ions and molecules with iron disulfide (14,15).

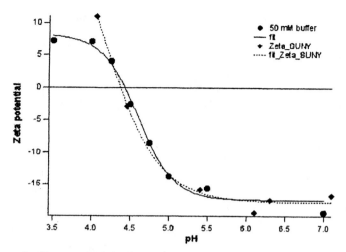

Figure 1: Zeta potential of mineralized ferritin measured by AC phase analysis light scattering under low and high ionic strength conditions.

References

1. Kim, I., Hosein, H.-A., Strongin, D. R., Douglas, T. Chemistry of Materials 2002, 14, 4874.
2. Wong, K. K. W., Colfen, H., Whilton, N. T., Douglas, T., Mann, S. J. Inorganic Biochemistry 1999, 76, 187.
3. Meldrum, F. C., Wade, V. J., Nimmo, D. L., Heywood, B. R., Mann, S. Nature 1991, 349, 684.
4. Meldrum, F. C., Heywood, B. R., Mann, S. Science 1992, 257, 522.
5. Meldrum, F. C., Douglas, T., Levi, S., Arosio, P., Mann, S. J. Inorg. Biochem. 1995, 58, 59.
6. Douglas, T., Stark, V. T. Inorg. Chem. 2000, 39, 1828.
7. Morel, F. M. M., Gschwend, P. M., "The role of colloids in the partitioning of solutes in natural waters," 1987.
8. Harrison, P. M., Arosio, P. Biochimica et Biophysica Acta 1996, 1275, 161.
9. Chasteen, N. D., Harrison, P. M. J. Struct. Biol. 1999, 126, 182.
10 Douglas, T., Ripoll, D. Protein Science 1998, 7, 1083.
11.Otsuka, S., Listowsky, I., Niitsu, Y., Urushizaki, I. J. Biol. Chem. 1980, 255, 6234.
12. Finsy, R. Advances in Colloid and Interface Science 1994, 52, 79.
13. Tscharnuter, W. W. Applied Optics 2001, 40, 3995.
14. J. Bebie and M.A.A. Schoonen, Geochemical Transactions 2000, 47.
15. J. Bebié, M.A.A. Schoonen, D.R. Strongin and M. Fuhrmann, Geochimica Cosmochimica Acta 624., 633-642, 1998.

Chapter 31

A Bioengineering Approach to the Production of Metal and Metal Oxide Nanoparticles

Hazel-Ann Hosein[1], Daniel R. Strongin[1], Trevor Douglas[2], and Kevin Rosso[3]

[1]Department of Chemistry, Temple University, Philadelphia, PA 19122
[2]Department of Chemistry and Biochemistry, Montana State University, 108 Gaines Hall, Bozeman, MT 59715
[3]Chemical Sciences Division, Pacific Northwest National Laboratory, P.O. Box 999, K8–96, Richland, WA 99352

Research is presented that explored the biological protein ferritin as a precursor for the production of nano-sized iron oxyhydroxide and iron particles supported on SiO_2. The protein, which is an iron sequestration protein, has a shell that is roughly spherical with a 120 Å outer diameter and a 80 Å internal diameter. Approximately 80 Å diameter ferrihydrite (i.e., Fe(O)OH) particles were grown within the shell in solution in our laboratory. The ferritin was deposited on a SiO_2 support, dried and subsequently exposed to oxygen in the presence of UV (i.e., ozone cleaning) that removed the protein shell. Atomic Force Microscopy (AFM) showed that these nano iron oxyhydroxide particles had a relatively narrow size distribution (averaging about 80 Å). Reduction of the iron oxyhydroxide nanoparticles in H_2 at elevated temperatures led to the production of iron metal nanoparticles, but with a broader size distribution than the iron oxyhydroxide precursor.

Introduction

Research in our laboratory is concentrated in part on developing a bio-mediated route to the synthesis and control of selected metallic and metal oxide nanoparticles, and to investigate the fundamental surface chemistry of these nano-structures. Our working hypothesis is that metallic nanoparticles will provide a unique chemistry not accessible at larger spatial dimensions. Much of the interest in nanoparticles stems from research that has shown that materials fabricated at such a size-scale often exhibit unique structural (both geometric and electronic) properties and chemical reactivity. Cobalt at the nanoscale, for example, shows a magnetic behavior that is dependent on size (1). Nano-sized TiO_2 shows specific enhancements in its redox properties (2), while nano-clusters of gold have been shown to exhibit a catalytic activity that cannot be duplicated by bulk gold particles (3-6). Au on TiO_2 is an interesting case of how unique catalytic activity can be obtained if the spatial dimension of the catalytic particle is brought into the nano-regime. Another contribution in this book by Rodriguez demonstrated that nanogold on TiO_2 shows a unique activity toward S-O bond cleavage while prior studies have shown that nanoparticles of Au supported on the TiO_2 surface exhibit a high activity toward CO oxidation: a chemistry that is not experimentally observable on bulk Au (3,4,6-9). Some of the physical properties of this system bring forward interesting general scientific challenges that need to be met in order to evaluate the full potential of nanoparticle driven catalysis. At elevated temperature and/or reactant pressures, where catalysis occurs, agglomeration of the Au nanoparticles is observed (10). Hence, methods to prevent the destruction of the active nanoparticle component of the catalytic surface need to be developed. The electronic structure (11), and ultimately the reactivity of Au nanoparticles show a strong dependence on size. Au particles with a diameter of about 2 nm show the largest CO oxidation behavior (10). This size dependence might be expected to occur in other metallic systems. This result emphasizes the need to synthesize nanoparticles with well defined sizes, so that surface reactivity can be correlated to size. Hence, general synthetic routes to the formation of homogeneous metallic nanoparticles on a variety of supports are needed.

The general synthetic approach outlined in this contribution is to use a biologically relevant protein, ferritin, to mediate the growth of homogeneous iron oxyhydroxide nanoparticles and to reduce the oxide to metallic iron. Ferritin is an iron storage protein found in many biological systems. Horse spleen ferritin (used in the current research), for example, is comprised of 24 structurally similar polypeptide units that self-assemble. The resulting spherical cage has an outer and inner diameter of 120 and 80 Å, respectively. The protein shell of ferritin is permeable to iron which can be readily added to or removed from the protein interior under the appropriate redox conditions. Up to about 3000 Fe atoms can be mineralized as nanoparticles of ferrihydrite [Fe(O)OH] *(12)*.

SYNTHETIC STRATEGY AND RESULTS

The general synthetic scheme for the nanometal production is depicted in Figure 1. Homogeneous nanoparticles of Fe(O)OH were readily synthesized by

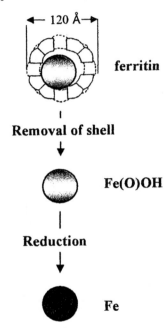

Figure 1: General synthetic scheme for Fe metal production.

methods described elsewhere *(13)*. Briefly, ferritin used in this study had an Fe loading of 1500. Deaerated solutions of $(NH_4)_2Fe(SO_4).6H_2O$ [25mM] were added to 2.5×10^{-6} M apoferritin (Sigma) in 20 ml MES buffer at pH 6.5 in three equivalent aliquots (0.2 ml each). After each addition the solution was allowed to air oxidize for 1 hour. The ferritin was dialyzed to remove buffer and approximately 0.2 ml of ferritin solution was deposited on to a SiO_2 wafer. The solution was subsequently dried in an anoxic environment.

After drying the sample was placed in an ozone cleaner (NovaScan) with 1 atm O_2. After heating the sample to 350 K for 1 hr the sample was removed and analyzed in an ultrahigh vacuum chamber using x-ray photoelectron spectroscopy (XPS). Figure 2 exhibits, N 1s and C 1s XPS data before and after cleaning. The N 1s level is absent after cleaning and the C 1s signal is markedly reduced. The remaining C 1s signal is thought to be primarily due to residual carbon that was present on the SiO_2 wafer prior to ferritin deposition [see spectrum (c)]. These data are taken to suggest that the protein shell is removed from the sample as presumably NO_x and CO_x gaseous species.

Atomic force microscopy (AFM) was used to characterize the sample morphology after ozone cleaning. Samples were delivered without atmospheric control from Temple University to Pacific Northwest National Laboratory for AFM analysis. AFM was performed in air under ambient conditions using a

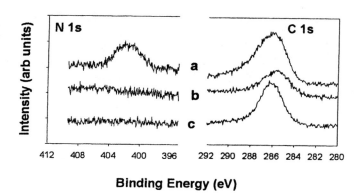

Figure 2: N 1s and C 1s XPS data for 1500 Fe loaded Ferritin before(a) and after(b) ozone cleaning. Spectrum (c) shows the residual carbon on the Si wafer prior to ferritin deposition.

Digital Instruments, Inc. BioScopeTM operating in Tapping mode^{TM}. Standard etched silicon probes (TESP) were used with resonant frequencies in the 250 Hz range and nominal tip radii of curvature of 5 -10 nm. The scan rate was 1.0 Hz. Figure 3a exhibits one such AFM image. We chose to base our particle size analysis of the image on the height field alone because isolated particles could not be found and the small size of the particles makes width analyses susceptible to overestimation due to tip curvature convolution effects. Figure 3a and the accompanying cross-section show the full range of height values to be 16.8 nm, with a root-mean-square (RMS) roughness (standard deviation of the height about the average value) of 2.13 nm. The peak-to-valley height differences for the large features in the cross-section are in the 8-9 nm range, consistent with the particle size range expected of the inorganic core of ferritin. Prior research showed that monodispersed films of the ferritin core could be obtained by heating the supported ferritin to 673 K in a nitrogen atmosphere *(14)*. Experimental observations in our laboratory suggest that the ozone cleaning more effectively removes residual carbon and nitrogen from the core and facilitates the reduction of the core when heated in hydrogen.

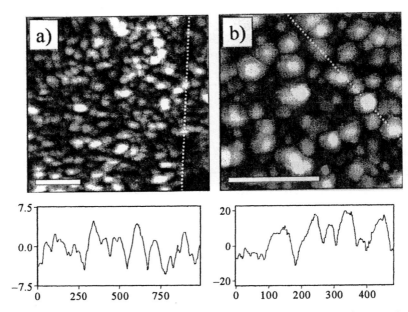

Figure 3: Tapping mde AFM height data for the a) iron oxyhydroxide, and b) reduced samples collected in air under ambient conditions. The scale bar in both images represents 250 nm. Dashed white lines indicate the locations for cross-sections shown below each image; the cross-section data are in nanometers. The height ranges for each image are given in the text.

Reduction of the metal oxide nanoparticles to the respective metal was carried out in a reaction cell intimately connected to the UHV chamber. Specifically, the protein free metal oxide particles were heated to 573 K in 50 Torr H_2 for 1 hr. The sample was then transferred into UHV where XPS was carried out. Figure 4 exhibits Fe 2p data and shows that there is a shift of the Fe $2p_{3/2}$ level from 711 eV (characteristic of Fe^{3+}-bearing oxides) to 707 eV that is characteristic of iron metal *(15)*.

Figure 3b exhibits AFM of the reduced sample. It was necessary, however, to expose the sample to air prior to acquiring the image. We assume that the AFM image of sample after exposure to air in Figure 3b is representative of the metal particles. The full range of height values in Figure 3b is 40.3 nm, with a

RMS roughness of 6.68 nm. Peak-to-valley height differences for the large features in the cross-section for the reduced sample are in the 19-20 nm range. Thus, assuming features in the images to be individual particles, the particles comprising the reduced sample were found to be considerably larger than those before reduction. Hence, under the experimental conditions used in this study, heating in H_2 had significant effects on the particle size distribution. Presumably, upon reduction neighboring particles underwent agglomeration. Based on the AFM observations, it is also possible that agglomeration was limited to the uppermost particles, as particles deeper in the aggregate overlayer appeared to be generally smaller in size (Fig. 3b). This would be consistent with the uppermost particles having the most degrees of freedom of motion for interaction with neighboring particles. This result for these initial samples was not completely unexpected, since the particles were deposited on the SiO_2 wafer as a multilayer structure where particles were in contact with each other. Prior research has shown that heating ferritin to 1173 K in a low vacuum results in the formation of roughly spherical fullerenic shells that surround individual iron nanoparticles (16). The majority of the particles in this prior study have a size below 13nm. Presumably, the shell around the iron particles decreased the amount of agglomeration, relative to our circumstance where the iron particles have no protective shell. Experiments in our laboratory at present are investigating the formation of true monolayer and submonolayer supported structures that will decrease agglomeration when subjected to our preparatory method.

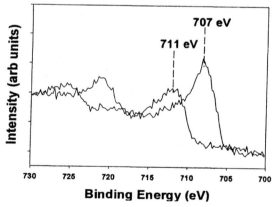

Figure 4: Fe 2p XPS data showing the shift of the Fe 2p3/2 level from 711 eV (characteristic of Fe^{3+}-bearing oxides) to 707 eV

SUMMARY AND GENERAL COMMENTS

Initial results from a bioengineering approach using ferritin as a precursor for supported Fe nanoparticle synthesis has been outlined. While relatively monodispersed iron oxyhydroxide particles were produced by this method, our initial results show that the metal particles undergo agglomeration reactions. We feel that this agglomeration can be limited by preparing true supported (sub)monolayer ferritin films that has been carried out by others (14-17). Ongoing work in our laboratory is investigating ways to prepare initial supported ferritin films in the monolayer and submonolayer regime. It is expected that this will yield isolated nanoparticles after protein removal and reduction, and will lessen any agglomeration effects relative to the multilayer deposition outlined in this contribution. Furthermore, experiments outlined in this contribution described ferritin that was loaded with Fe to its internal volume capacity. The iron loading of ferritin, however, can be varied continuously between 100 and 1000 iron atoms per ferritin cage by controlling the Fe(II):ferritin ratio (13,18). This is expected to give a range of nanoparticle sizes in our future experiments. Finally, the use of ferritin is not limited to iron containing nanoparticles, since prior research has shown that ferritin can also be filled with $Mn(O)OH$ (19) and $Co(O)OH$ (20).

Acknowledgments

DRS and TD acknowledge the Environmental Protection Agency (EPA) and the Donors of the Petroleum Research Fund (PRF), administered by the American Chemical Society, for support of this research. The AFM work described in this manuscript was performed at the W. R. Wiley Environmental Molecular Sciences Laboratory, a national scientific user facility sponsored by the U.S. Department of Energy's Office of Biological and Environmental Research and located at Pacific Northwest National Laboratory. PNNL is operated for the Department of Energy by Battelle.

References

1. Puntes, V. F.; Krishnan, K. M.; Alivisatos, A. P. *Science* **2001**, *291*, 2115-2117.
2. Hoffmann, M. R.; Martin, S. T.; Choi, W.; Bahneman, D. W. *Chemical Reviews* **1995**, *95*, 69-96.

3. Haruta, M.; Tsubota, S.; Kobayashi, T.; Kageyama, H.; Genet, M. J.; Delmon, B. *J. Catal.* **1993**, *144*, 175-192.

4. Valden, M.; Goodman, D. W. *Isr. J. Chem.* **1998**, *38*, 285-292.

5. Haruta, M.; Uphade, B. S.; Tsubota, S.; Miyamoto, A. *Res. Chem. Intermed.* **1998**, *24*, 329-336.

6. Haruta, M. *Stud. Surf. Sci. Catal.* **1997**, *110*, 123-134.

7. Takaoka, G. H.; Hamano, T.; Fukushima, K.; Jiro, M.; Yamada, I. *Nucl. Instrum. Methods Phys. Res., Sect. B* **1997**, *121*, 503-506.

8. Bondzie, V. A.; Parker, S. C.; Campbell, C. T. *Catal. Lett.* **1999**, *63*, 143-151.

9. Haruta, M.; Tsubota, S.; Kobayashi, T.; Ueda, A.; Sakurai, H.; Ando, M. *Stud. Surf. Sci. Catal.* **1993**, *75*, 2657-2660.

10. Chusuei, C. C.; Lai, X.; Luo, K.; Goodman, D. W. *Top. Catal.* **2001**, *14*, 71-83.

11. Lai, X.; Clair, T. P. S.; Valden, M.; Goodman, D. W. *Prog. Surf. Sci.* **1998**, *59*, 25-52.

12. Massover, W. H. *Micron* **1993**, *24*, 389-437.

13. Gider, S.; Awschalom, D. D.; Douglas, T.; Wong, K.; Mann, S.; Cain, G. *J. Appl. Phys.* **1996**, *79*, 5324-5326.

14. Yamashita, I. *Thin Solid Films* **2001**, *393*, 12-18.

15. Briggs, D.; Seah, M. P. *Practical surface analysis by auger and X-ray photoelectron spectroscopy*; John Wiley & Sons, Ltd.: New York, 1983.

16. Tsang, S. C.; Qiu, J.; Harris, P. J. F.; Fu, Q. J.; Zhang, N. *Chem. Phys. Lett.* **2000**, *322*, 553-560.

17. Furuno, T.; Sasabe, H.; Ikegami, A. *Ultramicroscopy* **1998**, *70*, 125-131.

18. Wong, K. K. W.; Douglas, T.; Gider, S.; Awschalom, D. D.; Mann, S. *Chem. Mater.* **1998**, *10*, 279-285.

19. Meldrum, F. C.; Douglas, T.; Levi, S.; Arosio, P.; Mann, S. *J. Inorg. Biochem.* **1995**, *58*, 59-68.

20. Douglas, T.; Stark, V. T. *Inorg. Chem.* **2000**, *39*, 1828-1830.

Chapter 32

Dendritic Nanoscale Chelating Agents: Synthesis, Characterization, and Environmental Applications

Mamadou S. Diallo[1,2,*], Lajos Balogh[3], Simone Christie[2], Piraba Swaminathan[2], Xiangyang Shi[3], William A. Goddard III[1], and James H. Johnson, Jr.[2]

[1]Materials and Process Simulation Center, Beckman Institute, MC 139–74, California Institute of Technology, Pasadena, CA 91125
[2]Department of Civil Engineering, Howard University, Washington DC 20015
[3]Center for Biologic Nanotechnology, University of Michigan, Ann Arbor, MI 48109–0355
*Corresponding author: diallo@wag.caltech.edu and mdiallo@howard.edu

Introduction

The complexation of metal ions is an acid-base reaction that depends on several parameters including (i) metal ion size and acidity, (ii) ligand molecular architecture and basicity and (iii) solution physicochemical conditions (1). Although macrocyles and their "open chain" analogues (unidentate and polydentate ligands) have been shown to form stable complexes with a variety of metal ions (1), their limited binding capacity (i.e. 1:1 complexes in most cases) is a major impediment to their utilization as high capacity chelating agents for industrial and environmental separations. Their relatively low molecular weights also preclude their effective recovery from wastewater by low cost membrane-based techniques (e.g., ultrafiltration). Moreover, their "uncontrolled" molecular composition, shape and size adversely impact their utilizations as templates for the synthesis of metal-bearing nanostructures with tunable electronic, magnetic, optical and catalytic activity. Recent advances in macromolecular chemistry such as the invention of dendritic polymers (2-3) are providing unprecedented opportunities to develop high capacity nanoscale chelating agents with well-defined molecular

238

composition, size and shape. Dendrimers are relatively monodisperse and highly branched nanoparticles with controlled composition and architecture consisting of three components: a *core*, *interior branch cells* and *terminal branch cells* (2-3). These nanoparticles can be designed to encapsulate metal ions and metal clusters. The sequestered metal ions and clusters can be toxic metal ions such as Cu(II), optically active metal ions such as Ag(I) and metal ions with catalytic properties such as Pd(II) (4-6). This research explores the fundamental science of metal ion uptake by dendritic nanoscale chelating in aqueous solutions and assesses the extent to which this fundamental knowledge could be used to develop: (i) high capacity and reusable chelating agents for environmental separations and (ii) redox and catalytically active nanoparticles with enhanced reactivity, selectivity and longevity for environmental detoxification.

Experimental Methods and Procedures

Dendrimer Synthesis and Characterization

We initially focused on the evaluation of poly(amidoamine) (PAMAM) dendrimers with ethylene diamine (EDA) core and terminal NH_2 groups (Figure 1). These dendrimers possess functional nitrogen and amide groups arranged in regular "branched upon branched" patterns, which are displayed in geometrically progressive numbers as a function of generation level. This high density of nitrogen ligands enclosed within a nanoscale container along with the possibility of attaching various functional groups such as carboxyl, hydroxyl, etc to PAMAM dendrimers (Figure 1) make them particularly attractive as high capacity chelating agents. Generations G3 and G4 PAMAM dendrimers with EDA core and terminal NH_2 groups were purchased from Dendritech (Midland, MI). G5 PAMAM dendrimer with terminal NH_2 groups, G4-COOH, G4-OH and G4-Ac PAMAM dendrimers with terminal COOH, OH and $NHCOCH_3$ terminal groups were synthesized at the University of Michigan Center for Biologic Nanotechnology. All the dendrimers were characterized by HPLC, ^1H and ^{13}C NMR spectroscopy, polyacryl amide gel electrophoresis (PAGE), size exclusion chromatography (SEC), capillary electrophoresis (CE), and matrix-assisted laser desorption (MALDI) mass spectrometry.

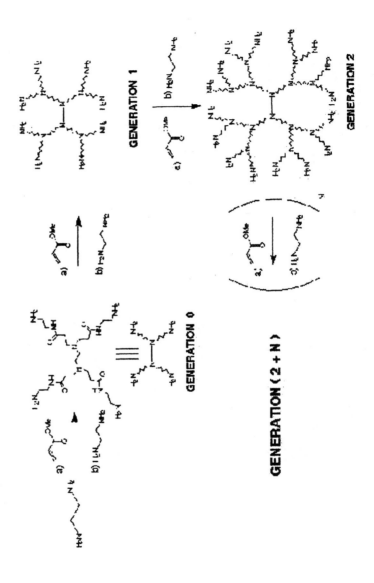

GENERATION 1

GENERATION 2

GENERATION 0

GENERATION (2 + N)

G4   Z =-NH₂

G4-COOH   Z =-COOH

G4-OH   Z =-OH

G4-Ac   Z =-NCOCH₃

Figure 1: Structures of EDA Core PAMAM Dendrimers with Terminal NH2, COOH, OH and NCHOCH3 Groups Evaluated in This Study (7)

Bench Scale Measurements of Proton and Metal Ion Bending by Dendrimers in Aqueous Solutions

Acid-base titration experiments of the PAMAM dendrimers evaluated in this study were performed in the Department of Civil Engineering at Howard University (HU). All titrations experiments were carried out using a computer-controlled QC-Titrate system from Man-Tech Associates and reagent grade HCl and NaOH as titrants. Cu(II) was selected as model cation for metal ions with affinity for N donors. Bench scale measurements were carried out at HU to assess the effects of metal-ion dendrimer loading, dendrimer generation and terminal group chemistry, and solution pH on the extent of binding (EOB) of Cu(II) in aqueous solutions of PAMAM dendrimers. The binding assay procedure consisted of (i) mixing and equilibrating aqueous solutions of Cu(II) and PAMAM dendrimers, (ii) separating the metal ion laden dendrimers from the aqueous solutions by ultrafiltration and (iii) and measuring the Cu concentrations of the equilibrated solutions and filtrates by atomic absorption spectroscopy (4). The concentrations of copper in the equilibrated tubes and filtrates were, respectively, taken as the initial concentration (Cu_0) and the concentration of unbound copper (Cu_a) in the aqueous solutions. The concentration of copper bound to a dendrimer (Cu_b) [mole/L] was expressed as:

$$Cu_b = Cu_0 - Cu_a \qquad\qquad 1$$

The extent of binding (EOB) [i.e., number of moles of Cu(II) bound per mole of dendrimer] was expressed as:

$$EOB = \frac{Cu_b}{C_d} \qquad\qquad 2$$

where C_d (mole/L) is the total concentration of dendrimer in the aqueous solution.

Results and Discussion

Overall, all the dendrimers synthesized in this study exhibit molar mass within 90-95% of the theoretically estimated molar mass of the corresponding "monodisperse" dendrimers shown in Figure 1 (7). Figure 2 highlights the effects of metal-ion dendrimer loading, dendrimer generation and solution pH on the EOB of Cu(II) in aqueous solutions of EDA core PAMAM dendrimers with terminal NH_2 groups. Not surprisingly, it is strongly affected by solution pH. In aqueous solutions of G3, G4 and G5 EDA core PAMAM dendrimers

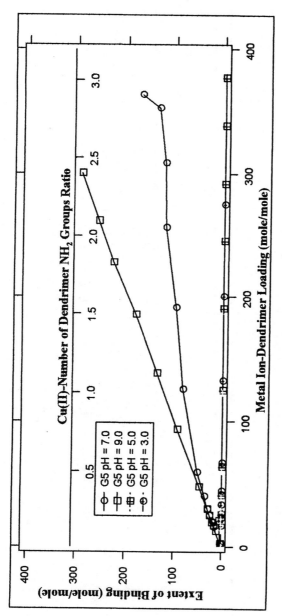

Figure 2: Extent of Binding of Cu(II) in Aqueous Solutions of G3, G4 and G5 EDA Core PAMAM Dendrimers with Terminal NH₂ groups as a Function of Solution pH and Metal Ion Dendrimer Loading at Room Temperature (7)

Continued on next page.

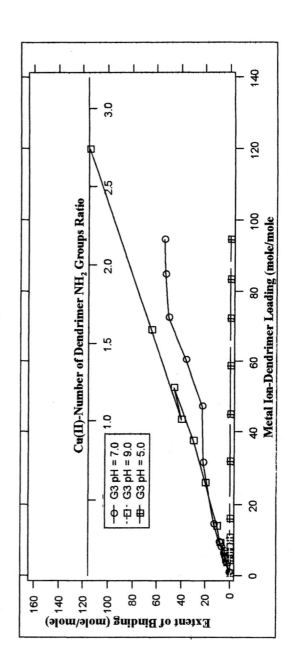

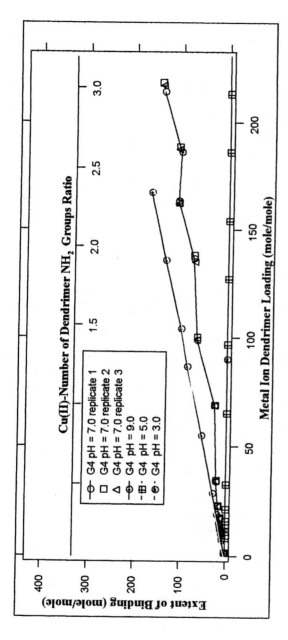

Figure 2. *Continued.*

with terminal NH_2 groups at pH = 9.0, all the dendrimer N groups are unprotonated (7). In this case, the EOB of Cu(II) increases linearly with metal ion-dendrimer loading for all the dendrimers. 100% of the Cu(II) ions are bound to the dendrimers in all cases. Conversely, no Cu(II) ions are bound to the dendrimers when all their terminal NH_2 and tertiary N groups become completely protonated at pH = 5.0 and 3.0 (7). A more complex metal ion uptake behavior is observed when the dendrimer N groups become partially protonated at pH = 7.0 (7). For G3 and G4 PAMAM dendrimers, the EOB of Cu(II) go through a series of 2 to 3 distinct binding steps as metal ion dendrimer loading increases. We attribute this behavior to the presence of sites with different Cu(II) binding affinity within the dendrimers. We also found that the EOB of Cu(II) increases linearly in aqueous solutions of G4-COOH, G4-OH and G4-Ac PAMAM dendrimers at pH = 9.0 (7.0). It also go through a series of 2 to 3 distinct binding steps as observed for G3 and G4 EDA core PAMAM dendrimers with terminal NH_2 groups at pH= 7.0. Compared to polydendate chelating agents and macrocycles with N donors such as triethylene tetramine and cyclams (1), our measurements show that dendrimers such EDA core PAMAM behave more like nanoscale containers for Cu(II) at pH = 7.0 and 9.0. A more detailed discussion of these results will be provided elsewhere (7).

Acknowledgements

This research was funded by the National Science Foundation (CTS Grant 0086727), the Environmental Protection Agency (STAR Grant R829626) and the Department of Energy HBCU/MI Environmental Technology Consortium (Cooperative Agreement DE/FC02-02EW15254).

References

Hancook, R. D. and Martell, A. E. *Metal Ion Complexation in Aqueous Solutions,* Plenum Press, New York, 1996.

Tomalia, D.A.; Naylor, A.M.; Goddard III, W. A. "Starburst® Dendrimers: Molecular-Level Control of Size, Shape, Surface, Chemistry, Topology, and Flexibility from Atoms to Macroscopic Matter". *Angew. Chem.*, **1990**, 102, 119.

Newkome, G. R.; Moorefield, C. N. and Vogtle, F. *Dendritic Molecules. Concepts-Syntheses-Perspectives*, VCH, New York, 1996.

Diallo, M. S.; Shafagati, A.; Johnson, J. H.; Balogh, L.; Tomalia, D.; Goddard, W. A., III, Poly(amidoamine) Dendrimers: A New Class of High Capacity Chelating Agents for Cu(II) Ions, .*Environ. Sci. and Technol.*, **1999**, 33, 820.

Balogh, L; Valluzzi, R.; Hagnauer, G. L.; Laverdure, K. S.;. Gidoand, S. P. and Tomalia, D. A. Formation of Silver and Gold Dendrimer Nanocomposites; *J. of Nanoparticle Res.*, 1999, 1, 353.

Brinkman, N.; Giebel, D.; Lohmer, M.; Reetz, M. T.; and Kragi, U. 'Allylic Substitution with Dendritic Palladium Catalysts in a Continuously Operating Membrane Reactor". *J. Catalysis*, **1999**, 183, 163.

Diallo, M.S. Balogh, L.; Christie, S.; Swaminathan, P.; Shi. X.; Wooyoung, U.; Papelis. L.; Johnson, J. H. Jr. and Goddard, W. A. III. Dendritic Chelating Agents 1. Uptake of Cu(II) Ions by Poly(amidoamine) Dendrimers in Aqueous Solutions. To Submitted to *Langmuir* in May 2003.

Chapter 33

Iron Nanoparticles for Site Remediation

Wei-xian Zhang, Jiasheng Cao, and Daniel Elliott

Department of Civil and Environmental Engineering, Lehigh University, Bethlehem, PA 18015

ABSTRACT

Nanoscale iron particles represent a new generation of environmental remediation technologies that could provide cost-effective solutions to some of the most challenging environmental cleanup problems. Nanoscale iron particles have large surface areas and high surface activity. More important, they provide enormous flexibility for in situ applications. Research at Lehigh University has demonstrated that nanoscale iron particles are very effective for the transformation and detoxification of a wide variety of common environmental contaminants, such as chlorinated organic solvents, organochlorine pesticides and PCBs. Modified iron nanoparticles, such as catalyzed and supported nanoparticles have been synthesized to further enhance the speed and efficiency of remediation. In this article, recent developments are assessed, including: (1) synthesis of nanoscale iron particles (<100 nm) from common precursors such as Fe(II) and Fe(III); (2) reactivity of the nanoparticles towards contaminants in soil and water, and (3) reactions of the nanoparticles in the subsurface.

Two factors contribute to the nanoparticles' capabilities as an versatile remediation method. The first is their small particle sizes (1-100 nm). In comparison, a typical bacterial cell has a diameter on the order of 1 micrometer (1,000 nm). Nanoparticles therefore have better access to contaminants sorbed into soils than typical baterial cells. Transport andmovement of nanosized particles are largely contolled by Brownina motion, not by gravity. Therefore, even the slow flow of groundwater might be sufficient to transport the nanosized particles. The nanoparticle-water slurry can be injected under pressure and/or by gravity to the contaminated plume where treatment is needed. Nanoparticles can also remain in suspension for extended periods of time to establish an in situ treatment zone. Nanoparticles also provide enormous flexibility for in situ applications. For example, nanoparticles are easily deployed in slurry reactors for the treatment of contaminated soils, sediments, and solid wastes. Alternatively, nanoparticles can be anchored onto a solid matrix such as activated carbon and/or zeolite for enhanced treatment of water, wastewater, or gaseous process streams. Examples of potential applications of nanoscale iron particles for site remediation are further illustrated in Figure 1.

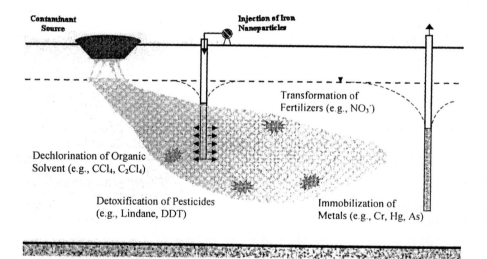

Figure 1: Nanoscale iron particles for in situ remediation. Recent research has suggested that as a remediation technique, nanoscale iron particles have several advantages: (1) effective for the transformation of a large variety of environmental contaminants, (2) inexpensive, and (3) nontoxic.

NANOSCALE IRON PARTICLES

Metallic or zero-valent iron (Fe^0) is a moderate reducing reagent, which can react with dissolved oxygen and to some extent with water:

$$2Fe^0_{(s)} + 4H^+_{(aq)} + O_{2(aq)} \rightarrow 2Fe^{2+}_{(aq)} + 2H_2O_{(l)} \tag{1}$$

$$Fe^0_{(s)} + 2H_2O_{(aq)} \rightarrow Fe^{2+}_{(aq)} + H_{2\,(g)} + 2OH^-_{(aq)} \tag{2}$$

The above equations represent the classical electrochemical/corrosion reactions by which iron is oxidized from exposure to oxygen and water. The corrosion reactions can be accelerated or inhibited by manipulating the solution chemistry and/or solid (metal) composition. Environmental chemistry of metallic or zero-valent iron has received tremendous attention since early 1990s when research established that oxidation of iron can be coupled to contaminant reduction. Contaminants such as tetrachloroethene (C_2Cl_4), a common solvent, can readily accept the electrons from iron oxidation and be reduced to ethene in accordance with the following stoichiometry:

$$C_2Cl_4 + 4Fe^0 + 4H^+ \rightarrow C_2H_4 + 4Fe^{2+} + 4Cl^- \tag{3}$$

Environmental applications of metallic iron have been enthusiastically accepted by users and regulatory agencies, largely due to the low costs and absence of any known toxicity induced by the use of iron. Use of metallic iron in the form of packed bed reactors and permeable reactive barriers has been widely reported.

Work at the authors' laboratory at Lehigh University has focused on the research and development of nanoscale iron particles for environmental remediation. We developed a method to synthesize nanoscale iron particles. Typically, nanoparticles can be prepared by using sodium borohydride as the key reductant. $NaBH_4$ (0.2 M) is added into $FeCl_3 \cdot 6H_2O$ (0.05 M) solution (~1:1 volume ratio). Ferric iron is reduced by the borohydride according to the following reaction:

$$4Fe^{3+} + 3BH_4^- + 9H_2O \rightarrow 4Fe^0\downarrow + 3H_2BO_3^- + 12H^+ + 6H_2 \tag{4}$$

Palladized Fe particles are prepared by soaking the freshly prepared nanoscale iron particles with an ethanol solution containing 1wt% of palladium acetate ($[Pd(C_2H_3O_2)_2]_3$). This causes the reduction and subsequent deposition of Pd on the Fe surface:

$$Pd^{2+} + Fe^0 \rightarrow Pd^0 + Fe^{2+} \tag{5}$$

Similar methods were used to prepare Fe/Pt, Fe/Ag, Fe/Ni, Fe/Co, Fe/Cu bimetallic particles. With the above methods, nanoparticles with average diameters in the range of 50-70 nm can be produced. Most particles (>80%) had diameters less than 100 nm with 30% less than 50 nm. The average specific surface area of the nanoscale Pd/Fe particles was about 35 ± 2.7 m^2/g. Ferrous iron salt (e.g., $FeSO_4$) as the precusor has also been successfully used.

ACTIVITY ASSESSMENT

Our research has established nanoscale iron particles as effective reductants and catalysts for a wide variety of common environmental contaminants including chlorinated organic compounds and metal ions. Examples are given in Table 1. For halogenated hydrocarbons, almost all can be reduced to benign hydrocarbons by the nano-Fe particles. Ample evidence indicates that iron based materials are effective for the transformation of many other contaminants, including anions (e.g., NO_3^-, $Cr_2O_7^{2-}$), heavy metals (e.g., Ni^{2+}, Hg^{2+}), and radionuclides (e.g., UO_2^{2+}).

In this section, results from a recent laboratory experiment are presented. The laboratory study was conducted as a part of a project to evaluate the potential of using nanoscale iron particles for in situ remediation of chlorinated organic solvents (e.g., 1,1,1-trichloroethane, trichloroethene) found in the soil and groundwater at an US Naval site. Groundwater and soil samples were collected at the site and were shipped to Lehigh University for various laboratory tests. The tests were conducted during the period of August 2002 to March 2003. Experiments were designed to determine the concentrations of organic contaminants, to investigate changes of groundwater chemistry as a result of the addition of the nano iron particles, and to examine the efficacy of the nanoparticles for dechlorination of major chlorinated organic compounds found in the groundwater and soil.

The groundwater contained approximately 6,070 μg/L of 1,1,1-trichloroethane (TCA), 4,680 μg/L of trichloroethene (TCE) and a few other chlorinated hydrocarbons (e.g., tetrachloroethene, dichloroethene) at lower concentrations (< 100 μg/L). Batch reactors were loaded with 80 mL water, 0 to 20 g soil, and nanoparticles at varied concentrations. At selected time intervals, a 5-10 μL headspace aliquot was withdrawn from the batch bottle for analysis.

Table 1: Common environmental contaminants that can be transformed by nanoscale iron particles.

Chlorinated Methanes	**Trihalomethanes**
Carbon tetrachloride (CCl_4)	Bromoform ($CHBr_3$)
Chloroform ($CHCl_3$)	Dibromochloromethane ($CHBr_2Cl$)
Dichloromethane (CH_2Cl_2)	Dichlorobromomethane ($CHBrCl_2$)
Chloromethane (CH_3Cl)	

Chlorinated Benzenes	**Chlorinated Ethenes**
Hexachlorobenzene (C_6Cl_6)	Tetrachloroethene (C_2Cl_4)
Pentachlorobenzene (C_6HCl_5)	Trichloroethene (C_2HCl_3)
Tetrachlorobenzenes ($C_6H_2Cl_4$)	cis-Dichloroethene ($C_2H_2Cl_2$)
Trichlorobenzenes ($C_6H_3Cl_3$)	trans-Dichloroethene ($C_2H_2Cl_2$)
Dichlorobenzenes ($C_6H_4Cl_2$)	1,1-Dichloroethene ($C_2H_2Cl_2$)
Chlorobenzene (C_6H_5Cl)	Vinyl Chloride (C_2H_3Cl)

Pesticides	**Other Polychlorinated Hydrocarbons**
DDT ($C_{14}H_9Cl_5$)	PCBs
Lindane ($C_6H_6Cl_6$)	Dioxins
	Pentachlorophenol (C_6HCl_5O)

Organic Dyes	**Other Organic Contaminants**
Orange II ($C_{16}H_{11}N_2NaO_4S$)	N-nitrosodimethylamine (NDMA)
Chrysoidine ($C_{12}H_{13}ClN_4$)	($C_4H_{10}N_2O$)
Tropaeolin O ($C_{12}H_9N_2NaO_5S$)	TNT ($C_7H_5N_3O_6$)
Acid Orange	
Acid Red	

Heavy Metal Ions	**Inorganic Anions**
Mercury (Hg^{2+})	Dichromate ($Cr_2O_7^{2-}$)
Nickel (Ni^{2+})	Arsenic (AsO_4^{3-})
Silver (Ag^+)	Perchlorate (ClO_4^-)
Cadmium (Cd^{2+})	Nitrate (NO_3^-)

Organic concentrations were measured by the static headspace gas chromatograph (GC) method. Concentrations of chlorinated volatile compounds were measured using a HP5890 GC equipped with a DB-624 capillary column (30m×0.32mm) and an electron capture detector (ECD). Hydrocarbon products in the headspace were qualitatively identified with a Shimadzu QP5000 GC-MS and further quantified with GC analysis by comparing retention times and peak areas with standard gas samples (ethane, ethene, acetylene, methane and carbon dioxide etc.).

It has been suggested that the nanoscale iron particles can be used as an efficient reducing reagent for regulating the standard potential in the subsurface and for chemical and/or biological dechlorination. According to Equations 1 and 2, iron-mediated reactions should produce a characteristic increase in pH and decline in solution redox potential (E_H). A highly reducing environment ($E_H <<$ 0) is created through the rapid consumption of oxygen and other potential oxidants and production of hydrogen. Typically in a closed batch reactor, a pH increase of 2-3 units was observed while ORP reduction was in the range of 500-900 mV. It is expected that the pH and ORP changes would be less dramatic for field applications wherein diffusion and other mechanisms dilute the chemical changes. Our previous field experiments suggest that the water pH was increased by about one pH unit, and ORP can be maintained at –300 mV to –500 mV. The pH increase and potential decrease at the site should favor the growth of anaerobic microorganisms, which could be beneficial for accelerated biodegradation. Production of hydrogen gas and also ferrous iron ion should further encourage microbial growth.

Rapid and complete dechlorination of all chlorinated contaminants was achieved within the water and soil-water slurries. For example, with a nanoscale Pd/Fe particle dose at 6.25 g/L, all chlorinated compounds were reduced to below detection limits (<10 µg/L) within eight hours (with and without soil). Ethane was the major product in all tests. Greater than 99% removal was achieved with nanoscale iron particle (no palladium) in 24 hours (Figure 2). Experiments with repeated spiking of TCE suggest that the laboratory-synthesized nanoscale particles can remain reactive within the soil and water environments for extended periods of time (6-8 weeks).

Further analysis of the batch test data suggests that the nanoscale Pd/Fe particles can achieve a dechlorination rate of ~1 mg TCE/g nano Fe/hour and a total capacity of approximately 100-200 mg TCE/g nano Fe. The rate and capacity parameters are approximately 1-3 orders of magnitude higher than conventional iron powders (e.g., >10 µm).

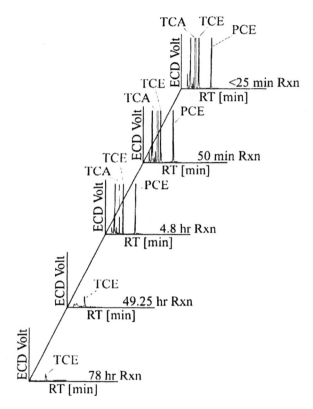

Figure 2.

CONCLUSIONS

Our recent work has demonstrated that: (1) nanocale iron particles (<100 nm) can be synthesized from inexpensive precursors such as Fe(II) and Fe(III) salts; (2) the iron nanoparticles are reactive towards awide range of contaminants in soil and water; (3) the iron nanoparticles can remain reactive for extended periods of time (weeks to months); furthermore (3) field tests have demonstrated that the injection of n anoparticle s lurry i nto a quifer i s r elatively uncomplicated resulting in minimal change in the aquifer permeability. Nonetheless, there are still many challenges facing this technology. More research should be directed to the nanochemistry in the environment, e.g., contaminant transformation at the nanoparticle-water interface. Field scale studies to investigate the transport, fate and reactions are clearly needed to determine the potential and limits of the technology.

Acknowledgements

Funding supports from National Science Foundation (NSF) and US Environmental Protection Agency (USEPA).

References

Glazier, Robert, Venkatakrishnan, R., Gheorghiu, F., Walata, L., Nash, R., Zhang, W. 2003. Civil Engineering, 73 (5): 64-69.

Zhang, W. 2003. Journal of Nanoparticle Research, Vol. 5, No. 3-4, 1-10.

Elliott, Daniel and Zhang, W. 2001. Environmental Science & Technology 35: 4922-4926.

Zhang, W., Wang, C., and Lien., H. 1998. Catalysis Today, 40, 387-395.

Wang, C., and Zhang, W. 1997. Environmental Science & Technology, 31(7): 2154-2156.

Chapter 34

Membrane-Based Nanostructured Metals for Reductive Degradation of Hazardous Organics

D. Meyer[1], D. Bhattacharyya[1,*], L. Bachas[2], and S. M. C. Ritchie[3]

Departments of [1]Chemical and Materials Engineering and [2]Chemistry,
University of Kentucky, Lexington, KY 40506
[3]Chemical Engineering, University of Alabama, Box 870203,
Tuscaloosa, AL 35487
*Corresponding author: db@engr.uky.edu

Introduction/Background

The technology development involving the remediation of waste waters containing chlorinated organic compounds has prompted a significant amount of research on understanding and improving the reaction kinetics involving zero-valent metals. In particular, the reduction of chlorinated ethanes and ethylenes by zero-valent iron (Fe^0) and modified bimetallic Fe^0 has received significant attention. The use of a second metal coating (such as Ni^0) reduces the reactivity loss of Fe^0 through oxidation. The initial problems with bulk zero-valent metal reduction included slow reaction rates and the presence of more toxic intermediates of dechlorination which were much harder to destroy [1]. Major improvements in reaction kinetics have already been demonstrated in literature using nanoscale metal particles [2, 3]. The surface-area normalized, pseudo-first order rate constant (k_{SA}) for the destruction of Trichloroethylene (TCE) by nanoscale Fe^0 is 3×10^{-3} $L \cdot m^{-2} \cdot hr^{-1}$ [2]. This result demonstrates a significant improvement over results for bulk (> 1 micrometer) particle dechlorination, where the value of k_{SA} is typically 6.3×10^{-5} $L \cdot m^{-2} \cdot hr^{-1}$ [4]. Further improvement using bimetallics has also been demonstrated in the literature with a reported k_{SA} of 0.098 $L \; m^{-2} \; h^{-1}$ [5].

For our work, membrane immobilization has been used to further increase the dehalogenation rates by using nanoscale zero-valent iron (Fe^0) and bimetallic Fe^0/Ni^0 for the reductive destruction of toxic organic compounds. The reasoning for this is that it provides two key advantages over traditional aqueous-phase systems. One advantage lies within the ability of the polymer to allow controlled growth of the zero-valent metal nanoparticles. The other advantage is related to local organic pollutant concentration around the Fe^0 nanoparticles. Selective partitioning by the organic should result in a higher localized concentration of organic (for example TCE) in the membrane domain and provide a significant enhancement of reaction rates. In addition, a polymer phase will have insignificant inorganic (for example, nitrate) partitioning, helping to minimize loss of Fe^0 to side reactions.

The use of polymeric matrices to facilitate nanoparticle growth has been well documented. The main advantage of polymer-based techniques is the enhanced ability to control the size distribution of the resulting nanoparticles. Through functionalization, polymeric matrices can be assembled in a way that provides limited space for nanoparticle growth as a result of steric hindrance within the matrix [6]. A typical method of formation is to prepare a dense membrane by phase-inversion from a homogeneous solution containing both a polymer and dissolved metallic salt. The resulting ions within the thin film are then reduced using a suitable reducing agent, such as sodium borohydride. A viable alternative to this approach involves thermolysis of metal-carbonyl compounds within a polymer film to produce the reduced metal nanoparticles [7]. The incorporation of multifunctional ion-exchange groups[8] in polymer matrix also provides a platform for metal capture and subsequent metal reduction.

Materials/Methods

Cellulose acetate has been selected for the membrane polymer, primarily because fabrication of the membrane can be accomplished using the easier phase-inversion techniques outlined above. Both Fe^0 and Fe^0/Ni^0 systems are investigated. The model compounds chosen for this study are trichloroethylene (TCE) and 1,1,2,2-tetrachloroethane (TtCA). Nickel chloride (crystalline, 99.6%), ferrous chloride (crystalline, certified), ferric nitrate (crystalline, 99.99%), ferrous sulfate (crystalline, 99%, A.C.S.) and ferric chloride (lumps, 99%) were obtained and used as sources of Fe^0 in nanoparticle synthesis. Cellulose acetate (Lot # A17A) was obtained from Kodak to be used as the membrane. $NaBH_4$ (powder) was used as the reducing agent for all cases. TtCA (98%) and TCE (99.9%, A.C.S.) were obtained and used, as is, in dechlorination studies. Acetone (99.5%, A.C.S.), pentane (99%, spectrophotometric), 1,2-dibromoethane (99%), and sodium nitrate (aq) were acquired for use in analytical techniques. All water used in the experiments was de-ionized ultra-filtered water. All chemicals, except where noted, were obtained from either Fisher Scientific or Aldrich.

Membranes were obtained by first preparing a desired metal solution (either Fe or Fe/Ni) using the salts mentioned above. This solution was then added to a casting solution containing cellulose acetate dissolved in acetone until the final composition by weight of the casting solution was typically 17% cellulose acetate, 10% water, 1.2% Fe^{m+} and the balance acetone. The solution was mixed and allowed to sit until the presence of gas bubbles within the solution could not be visibly detected. A film was cast on the glass plate using a Gardner's® knife to insure uniform thickness. Phase inversion was performed either by direct solvent evaporation or solvent exchange using a coagulation bath. Complete evaporation proceeded until the polymer film began to pull away from the plate. For wet phase-inversion, the glass plate was immersed in a water bath for 15 min to allow complete de-mixing of the polymer solution. Coagulation temperature (either 1°C or 21°C) was selected based on desired film properties. Metals ions were reduced using sodium borohydride. Depending on the type of phase inversion, the reduction occurred either during coagulation or as a separate step.

Batch dechlroination experiments for both TCE and TtCA were performed using the reduced membranes. A selected area of the membrane was added to 120 mL bottles equipped with Mininert® (Supelco) valves, containing 110 mL of a chlorinated compound with concentrations ranging from 10 to 100 ppm of the target compound. The bottles were shaken for the desired reaction time and sampled for organic analysis by PT-GC/MS, Cl⁻ (excluding cases where Chloride-based metallic salts were used), pH, and metal loss by AA. For selected cases, membranes were examined by either SEM or TEM equipped with EDX to determine particle size and qualitative content.

Results/Discussion

The typical films obtained using phase inversion were approximately 24 cm by 24 cm with a thickness of 100-150 μm. For cases where Fe was analyzed, total losses were less than 2 mg (6 % of the maximum). Typical membrane loading was around 40 mg in 2 g of cellulose acetate. Results from TEM/SEM analysis of the membranes are shown in Figure 1. The membrane-immobilized particles were found to have an average particle diameter from 25-30 nm to 240 nm, depending on the metal salt used. These particles are much better than clusters obtained in solution phase, which had a bimodal size

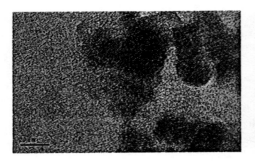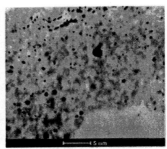

Figure 1: TEM and SEM photos of Fe/Ni and Fe in a cellulose acetate thin film.

distribution around 300 and 640 nm. The particle distribution shows more particles formed near the surfaces of the membrane. For the case of Fe/Ni, the metals were also very stable in air.

Films prepared using wet phase inversion were used in studies of TCE and TtCA dechlorination by both Fe^0 and Fe^0/Ni^0. A summary of results for the various membrane trials is given in Tables 1 and 2. The data given for parent compound degradation (expressed as C/C_O) have been corrected for adsorption by extracting the membrane in pentane. The amount of iron surface area is calculated as "g Metal per L of solution treated". For single point experiments, the value of C_O does not change. In multi-point experiments this is not the case since a portion of the TtCA was removed with each sample taken.

The membranes used for the experiments in Table 1 were ~ 25 cm x 25 cm x 100 μm, with the metal accounting for ~ 1-% by weight of the membrane. The data for TtCA destruction corresponds to a k_{SA} value of 0.17 L m^{-2} h^{-1},

Table 1: TtCA and TCE destruction using Fe^0 nanoparticles in cellulose acetate

	Reaction Time (hrs)	Fe^0 (mg L^{-1})	Parent Compound $C_W/C_{W,O}$	Cl$^-$ C_{Cl}/C_{max}
TtCA				
Fe-CA	2	70	0.5-0.8	0.1-0.2
TCE				
Fe-CA	1	420	0.59	.015

Table 2: TtCA and TCE destruction using Fe^0/Ni^0 (4:1 metal ratio) nanoparticles in cellulose acetate.

	Reaction Time (hrs)	Fe^0 (mg L^{-1})	Parent Compound C/C_O	k_{SA}* L m^{-2} h^{-1}
TCE				
	2	418	0.71-0.86	
	3.17	423	0.32	2.8E-02
	4.25	453	0.27	
	5	419	~ 0	
TtCA				
	2.5	427	0.51	
	16	429	0.01	

which is 1 order of magnitude better than the value 0f 0.012 L m^{-2} h^{-1} obtained by Gillham and coworkers using bulk Fe^0 [9]. The initial rate obtained for TCE is 0.12 L m^{-2} h^{-1}, which is 4 orders of magnitude better than the values for bulk Fe^0 (6.3×10^{-5} L m^{-2}hr^{-1} [4]) and 2 orders of magnitude better than values reported for nanoparticles (3×10^{-3} L m^{-2}hr^{-1} [2]). These values are significant since the amount of metal used was typically an order of magnitude lower than solution phase experiments.

The data obtained for Ni/Fe bimetallic systems was also obtained using cellulose acetate films ~ 25 cm x 25 cm x 100 μm. As indicated, the molar Fe-to-Ni ratio is 4:1. Ideally this ratio can be made much larger since Ni is considered a toxic metal and is not ideal for environmental application. However incorporation of a membrane could allow for potential recapture of metal ions to prevent release into the environment. For TCE, the parent compound was completely destroyed in 5 hrs. Headspace analysis showed ethane, butane, and hexane as the only products. This is consistent with literature findings [5] suggesting that ethane is polymerized at the Ni surface to form even numbered saturated alkanes. Shorter time points revealed the presence of dichloroethylene (both cis- and trans-), which is also consistent with proposed mechanisms for TCE destruction. The value of k$_{SA}$ determined for this system is 0.028 L m^{-2} h^{-1}, which is close to the value of 0.098 L m^{-2} h^{-1} obtained using Fe/Ni in solution with a 3:1 Fe-to-Ni ratio [5]. This data suggests that some membrane loading time is required before the reaction rate reaches its peak. However this effect could be eliminated by the use of a convective flow system to reduce mass transfer limitations for uptake of the chlorinated organic into the membrane. For TtCA, only initial results are available. It is worth noting that over 50-% destruction of TtCA occurs within the first 2 hours, with almost complete destruction being demonstrated within 16 hours. In this case, again ethane, butane, and hexane were observed to be the only apparent identifiable products. However lack of literature data will require further studies to determine k$_{SA}$, as well as the mechanism of destruction for this system.

It has been shown that Fe and Fe/Ni nanoparticles in the range of 25-30 nm can be synthesized in cellulose acetate films using phase inversion membrane fabrication techniques coupled with chemical reduction by sodium borohydride. The application of these films to dechlorination systems showed at least an order of magnitude improvement over solution phase kinetics using only a fraction of the metal required. The availability of a wide variety of specialty polymeric materials allows for a more versatile application of zero-valent metal reactions. In addition it allows for the possibility of metal recapture to prevent the release of toxic metals, i.e. Ni, into the environment during treatment. This is significant since bimetallic systems exhibit much faster kinetics with a more benign product distribution.

This work is made possible with funding provided by the EPA-STAR program, NSF-IGERT, and SBRP-NIEHS.

References

1. Arnold, W. A.; Roberts, L. A. *Environ. Sci. Technol.* **2000**, *34*, 1794-1805.
2. Wang, B. W. and W. X. Zhang. *Environ. Sci. Technol.* **1997**, *31*, 2154-2156.
3. Choe, S., *et al. Chemosphere* **2001**, *42*, 367-372.
4. Johnson, T. L.; Scherer, M. M.; Tratnyek, P. G. *Environ. Sci. Technol.* **1996**, *30*, 2634-2640.
5. Schrick, B.; Blough, J.; Jones, A.; Mallouk, T. *Chem. Mater.* **2002**, *14*, 5140-5147.
6. Sidrov, S. N.; Bronstein, L. M.; Davankov, V. A.; Tsyurupa, M. P.; Solodovnikov, S. P.; Valetsky, P. M.; Wilder, E. A.; Spontak, R. J. *Chem. Mater.* **1999**, *11*, 3210-3215.
7. Tannenbaum, R. *Curr. Trends Poly. Sci.* **1998**, *3*, 81-98.
8. Bhattacharyya, D.; Hestekin,J.; Brushaber,P.; Cullen, C.; Bachas,L.G.; and Sikdar,S. *J. Membrane Science.* **1998**,*141*, 121-135.
9. Gillham, R. W.; O'Hannesin, S. F. *Ground Water* **1994**, *32*, 958-967.

Chapter 35

Porous Membranes Containing Zero-Valent Iron Nanorods for Water Treatment

S. M. C. Ritchie[1,*], T. N. Shah[1], L. Wu[1], C. Claiborn[2], and J. C. Goodwin[1]

[1]Chemical Engineering Department and [2]Metallurgical and Materials Engineering Department, University of Alabama, Tuscaloosa, AL 35487

1. Introduction

Zero-valent iron has long been known to be very active for the dechlorination of chlorinated organics (1). Early studies showed the efficacy of granular iron for degradation of trichloroethylene (2,3). These studies showed that although the reduction was effective, the reaction was slow, and could result in significant formation of degradation by-products (3). Reducing the size of the iron particles to the nano-range has been very effective for increasing the rate of reaction, and subsequently reducing the formation of by-products (4). However, with increased activity came decreased material stability. Zero-valent iron nanoparticles are subject to oxidation in air and hydrolysis in water. In both cases, these reactions significantly reduce the efficiency of these materials for large-scale application. That is, how do you store the materials and keep them active until they are needed? In this work, composite, polymeric materials are formed containing zero-valent nanoparticles. The nanoparticles have been formed in-situ and ex-situ, although the final reduction reaction occurs in the membrane-phase. In this paper, the formation of porous membranes containing zero-valent iron nanorods is discussed, and preliminary results are given to demonstrate the promise of these materials for degradation of chlorinated organics.

2. Methods

Two methods were used to make nanoparticle containing membranes. In the first method, polystyrene grafts were grown in the pores of polyethersulfone membranes. The polystyrene grafts are formed by cationic polymerization from sulfonic acid sites in the membrane. The mechanism for polymerization does not allow branching of the polymer. The grafts are subsequently treated with sulfuric acid to created sulfonated polystyrene grafts. The grafts are used for adsorption of iron cations. The sorption step is followed by reduction of the iron with sodium borohydride. This results in iron nanorods immobilized in the pores of the polyethersulfone membrane. The nanorods are located in the flow paths of the membrane, so they may readily contact and react with chlorinated organics in a permeated aqueous solution.

The second method involved solution formation of the nanoparticles similar to Li et al (5). Ferric chloride was dissolved in a 1:1:8 solution of cetyltrimethylammonium bromide (CTAB), n-butanol, and cyclohexane. The mixture was reduced using a 2 wt% solution of sodium borohydride in water to make the nanoparticles. The particles were diafiltered successively with water and methanol to remove residual surfactant and chloride. The nanoparticles were stored in methanol until needed. The membrane casting mixture was formed by adding the methanol slurry to a solution of cellulose acetate (CA) in acetone. There is no precipitation of the cellulose acetate, so nanoparticles can be dispersed throughout the mixture. Membranes were formed by casting the mixture on a glass plate and phase-inversion in ethanol.

TCE degradation studies were performed in batch experiments. In a typical experiment, TCE was diluted to 0.75 - 2 mM with degassed water. The membrane was cut into pieces and was added to 40 mL of the TCE solution in a Teflon-cap sealed vial. Samples are allowed to react for up to 72 hours in a wrist-shaker. The membrane was removed from the reaction mixture prior to chloride ion analysis using an ion selective electrode. A material balance was used to determine maximum TCE degraded assuming complete dechlorination. TEM micrographs were obtained by application of the nanoparticle-methanol slurry to a TEM grid and subsequent evaporation of the alcohol.

3. Results and Discussion

In-Situ Formed Nanorod Dechlorination. Studies with the nanorods formed on sulfonated polystyrene grafted membranes showed only minimal dechlorination.

A summary of these studies is shown in Table I. In both cases, the maximum available iron was only 5 mg. This is significantly lower than for comparable bulk metal studies. Long term studies yielded similar results to short-term studies, indicating that the reaction may be limited by the amount of iron. The starting pH was also less than 6, promoting side reactions for hydrolysis of the iron nanoparticles. The importance of reaction pH has been shown previously by Chen et al (6). Future work on these materials will focus on increasing the amount of iron in the membranes, along with better characterization of side products.

Table I: TCE Degradation with In-Situ Formed Iron Nanorods

TCE (mM)	Max Fe^0 (mg)	Time (hr)	[Cl⁻] (mM)	% Conversion
1.5	5	45	0.182	4.1
1.5	5	12	0.158	3.5

Ex-Situ Formed Nanorod Dechlorination. Membranes made with ex-situ formed nanoparticles were much more effective for TCE dechlorination. A summary of results for these membranes, including comparisons with iron powder, is given in Table II. The importance of having an active surface is shown with the first two entries. Long-term studies with a non-acid washed powder showed no dechlorination. However, the acid-washed metal provided almost 20% formation of the maximum chloride. Results for nanorod containing membranes gave better conversion with only a fraction of the iron. There is still not complete degradation of the chlorinated organic, but it does show the usefulness of incorporating nanoparticles in a polymer film.

Table II: TCE Degradation with Granular Iron and Ex-Situ Formed Nanorods

TCE (mM)	Form Fe^0	Fe^0 (mg)	Time (hr)	[Cl⁻] (mM)	% Conversion
2	100 mesh powder	100	45	0	0 (same as control)
2	Acid-washed iron powder	100	48	1.08	18
0.75	Surfactant Solution Fe^0 in CA	21.5	29	1.04	47
0.75	Solution Fe^0 in CA	100	22	0.13	8

These membranes are very stable and are gray to black in color. There is little to no oxidation of the films, unlike the particles, during storage for up to 1 month. During the degradation studies, the membranes turn orange which is indicative of iron oxidation. TEM micrographs of the nanoparticles used in this research are shown in Figure 1. The particles are very small (<50 nm), and seem to align in nanometer width rods. Accessibility of the iron surface in this case is much more apparent than would be observed with the iron powder (> 150 μm) used for comparison.

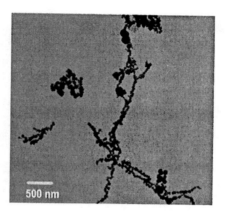

Figure 1: Nanorods of iron nanoparticles.

References

(1)   Gotpagar, J.; Grulke, E.; Tsang, T.; Bhattacharyya, D. Environ. Prog. 1997, 16, 137-143.

(2)   Gillham, R.W.; O'Hannesin, S.F. Ground Water 1994, 32, 959-967.

(3)   Matheson, L.J.; Tratnyek, P.G. Environ. Sci. Technol. 1994, 28, 2045-2053.

(4)   Lien, H.-L.; Zhang, W.-X. Coll. Surf. A: Physicochem. Eng. Aspects 2001, 191, 97-105.

(5)   Li, F.; Vipulanandan, C.; Mohanty, K.K. Coll. Surf. A: Physicochem. Eng. Aspects 2003, 223, 103-112.

(6)   Chem, J.-L.; Al-Abed, S.R.; Ryan, J.A.; Li, Z. K. Haz. Mat'ls. 2001, B83, 243-254.

Environmental Applications: Nanocatalysts for Environmental Technology

Editor
Sarah C. Larsen

Chapter 36

Nanocatalysts for Environmental Technology

Sarah C. Larsen

Department of Chemistry, University of Iowa, Iowa City, IA 52242

Introduction

Typical industrial heterogeneous catalysts are inorganic solids that consist of metal, metal oxide or metal sulfide nanometer-sized particles dispersed on high surface area supports, such as alumina.*(1-4)* These materials can be used to catalyze reactions such as hydrocarbon conversion reactions, partial oxidation reactions, and hydrodesulfurization reactions, to name just a few. Heterogeneous catalysts based on supported metal nanoparticles date back to the 1920's and therefore represent the earliest successful commercial applications of nanotechnology.*(2)* Many things have changed since the early 20th century, notably the scientific tools, such as electron and probe microscopies, that are currently available to investigate the fundamental properties and reactivity of nanometer-sized particles. Using these tools, scientists can now investigate, manipulate and control the surface reactivity of nanocatalyst materials on the atomic scale.

Nanostructured materials have important applications in environmental technology as environmental catalysts.*(5-8)* Environmental catalysis refers to catalytic technologies that address environmental issues, such as waste treatment and remediation, pollution prevention and the development of sustainable chemical processes.*(7,8)* One of the earliest success stories illustrating the use of a nanocatalyst for environmental technology is the catalytic converter that has been present in the exhaust manifold of automobiles since the early 1970's.*(1,9)* The catalyst in a catalytic converter consists of porous alumina which contains nanometer-sized particles of platinum, rhodium, ceria and zirconia. The role of

268

the platinum nanoparticles is to oxidize hydrocarbons and carbon monoxide and the function of the rhodium nanoparticles is to reduce NO_x. The ceria and zirconia components enhance the oxidation capabilities of the catalytic converter. The catalytic converter is an example of an environmental technology designed to reduce harmful emissions in a waste treatment process. Another category of environmental technologies is pollution prevention in which the goal is to completely eliminate waste production at the source. Applications of nanocatalyts to 1) waste treatment and remediation and 2) pollution prevention and waste minimization are discussed in more detail in the next two sections.

Waste Treatment and Remediation

Nanocatalyst materials have many important applications in waste treatment and remediation. Waste treatment and remediation includes applications such as air quality (indoor and outdoor), soil and water remediation, as well as emission controls for NOx and volatile organic compounds (VOC's). The goal of a nanocatalyst for these applications is to achieve high conversions and high selectivities so that the toxic pollutants are converted into more environmentally benign compounds. Several advantages can be realized using nanostructured catalyst materials for waste treatment and remediation processes.*(4)* Nanostructured catalyst materials exhibit different catalytic activities and selectivities based on their size, shape, composition and surface properties. By varying these structural and compositional parameters, nanocatalysts with specific properties can be designed for applications in waste treatment and remediation.

In this section, the recent work of Klabunde and coworkers on the use of oxide nanoparticles for the destructive adsorption of chemical toxins and toxic industrial chemicals is described. Klabunde and coworkers have successfully used nanoparticles of oxides, such as MgO, and mixed metal oxides to destructively adsorb toxic industial chemicals.*(10)* In other applications described in this chapter, nanostructured photocatalyst materials, such as TiO_2, have been used to degrade organic contaminants from polluted water and air. Current research by Shah and coworkers, focuses on increasing the photocatalytic efficiency in the visible range through the use of dopants, thereby changing the composition of the nanostructured materials.*(11)*

Pollution Prevention and Waste Minimization

Pollution prevention is defined as "source reduction and practices that efficiently use raw materials, energy, water or other resources to reduce or eliminate the creation of waste."*(12)* The concept of pollution prevention is at the heart of green chemistry. Specifically, green chemistry is concerned with the design of chemical products and processes that reduce or eliminate the production of toxic chemicals.*(13)* The holy grail in catalysis research is to achieve 100% selectivity in a catalytic reaction which is compatible with the green chemistry perspective.. In the 12 principles of green chemistry, Anastas states that "catalytic reagents should be as selective as possible"*(13)* By achieving 100% selectivity, the production of by-products which must be utilized or disposed will be avoided and the goal of pollution prevention will be achieved.

Nanocatalyst materials such as metal nanoclusters, nanocrystalline metal oxides, carbides or sulfides and nanostructured aluminosilicates are promising catalysts for environmental technologies designed to fulfill the goal of pollution prevention. For example, supported nanometer-sized clusters of gold exhibit very different catalytic reactivities relative to bulk gold. Supported gold nanocluster catalysts are selective catalysts for CO oxidation and other partial oxidation reactions.*(14)* Similarly, supported palladium nanoclusters have been investigated for CO and NO oxidation reactions as described by Ozensoy and coworkers in this section. Son and coworkers describe the use of nanometer-sized octahedral molecular sieves for applications to the environmentally friendly selective oxidation of hydrocarbons. Grassian and coworkers report the use and advantages of nanocrytalline zeolites as environmental catalysts. Future development of nanocatalyst materials with the goal of pollution prevention and waste minimization will lead to the implementation of sustainable processes and products.

Summary

Nanocatalyst materials hold significant promise for future applications in environmental technology. Potential nanocatalyst materials include metal nanoclusters, nanocrystalline metal oxides, carbides or sulfides, ceramics or composites, and nanostructured aluminosilicates. These nanoscale materials have many unique properties and, using the new tools of nanoscience, can be tailored for specific applications in order to achieve increased selectivity and yield for a catalytic process. Some of the advantages that may be achieved by improved nanocatalyst materials include increased energy efficiency and conversion, reductions in chemical waste, more effective waste remediation and the development of sustainable processes and products. The environmental benefits of nanocatalysts include cleaner air and water and, ultimately, a sustainable future.

References

1. Bell, A. T. *Science* **2003**, *299*, 1688-1691.
2. Somorjai, G. A.; Borodko, Y. G. *Catal. Lett.* **2001**, *76*, 1-5.
3. Gates, B. C. *Catalytic Chemistry*; Wiley: New York, **1992**.
4. Ying, J. Y. *AIChE Journal* **2000**, *46*, 1902-1906.
5. Armor, J. N. *Catal. Today* **1997**, *38*, 163-167.
6. Armor, J. N. *Applied Catalysis, A: General* **2000**, *194-195*, 3-11.
7. Armor, J. N. *Applied Catalysis B: Environ.* **1992**, *1*, 221-256.
8. Centi, G.; Ciambelli, P.; Perathoner, S.; Russo, P. *Catal. Today* **2002**, *75*, 3-15.
9. Bosch, H.; Janssen, F. *Catalysis Today* **1987**, *4*, 369-529.
10. Choudary, B. M.; Mulukutla, R. S.; Klabunde, K. J. *J. Am. Chem. Soc.* **2003**, *125*, 2020-2021.
11. Li, W.; Shah, S. I.; sung, M.; Huang, C.-P. *Journal of Vacuum Science and Technology* **2002**, *20*, 2303-2308.
12. Masciangioli, T.; Zhang, W.-X. *Environ. Sci. and Technol.-A Pages* **2003**, *37*, 102A-108A.
13. Anastas, P. T.; Williamson, T. C. In *Green Chemistry: Designing Chemistry for the Environment*; Anastas, P. T., Williamson, T. C., Eds.; American Chemical Society, **1996**, pp 1-17.
14. Choudhary, T. V.; Goodman, D. W. *Topics in Catal.* **2002**, *21*, 25-34.

Chapter 37

Nanocrystalline Metal Oxides: A New Family of Mesoporous Inorganic Materials Useful for Destructive Adsorption of Environmental Toxins

K. J. Klabunde*, G. Medine, A. Bedilo, P. Stoimenov, and D. Heroux

Department of Chemistry, Kansas State University, Manhattan, KS 66506
*Corresponding author: kenjk@ksu.edu

Introduction and Background

Although nanoscale materials promise to revolutionize many of our industries, including electronics, health care, energy and more, the near term uses are in environmental remediation and green chemistry applications.[1] One reason for this is that nanomaterials present unique properties as adsorbents and catalysts, because: (1) they possess high surface areas with large surface to bulk ratios so that the nanomaterial is used efficiently; (2) nanocrystals have unusual shapes and possess high surface concentrations of reactive edge, corner, and defect sites that impart intrinsically higher surface reactivities; (3) a wide range of Lewis acid/base properties and oxidation/reduction potential can be engineered into the nanomaterials since the periodic table of the elements (and their oxides, sulfides, etc.) becomes a literal playground for their design, and in a sense becomes three-dimensional since nanocrystal size is important as well as chemical potential; and (4) many nanocrystalline materials, especially ionic metal oxides, can be aggregated (pelletized) while still maintaining high surface areas and open pore structures.[2,3] Therefore, these nanomaterials represent a new family of porous, inorganic sorbent/catalyst materials perhaps as potentially useful as the zeolites[4] and other fascinating and useful materials, for example the MCM-41 silica series,[4] and the ETS-10 titania-silica zeolitic materials.[5]

Method and Materials

In this short abstract we present several examples where these new nanostructural materials have proven useful in environmental and catalytic applications. Generally, the nanomaterial in a powder form is allowed to contact a gaseous or liquid adsorbate. The rate of disappearance of the adsorbate is followed by GC-MS while the solid is monitored for changes by solid state NMR, IR, or Raman spectroscopies.

Results and Discussion

Destructive Adsorption of Chemical Toxins and Toxic Industrial Chemicals

Nanoparticles of MgO, CaO, SrO, Al_2O_3, and intimately mixed oxides $MgO-Al_2O_3$, and $SrO-Al_2O_3$ (all prepared by modified aerogel procedures-MAP) have proven effective in ambient temperature detoxification of chemical warfare agents (organophosphorus nerve agents and mustard), and their simulants (paraoxon, 2-chloroethylethylsulfide, organophosphorus fluorides and others).[6,7] The reactions of these liquid adsorbates with the dry powder nanomaterials are rapid upon contact, and further penetration into the fine powder is governed by material transfer by adsorbate vapor pressure. This material transfer step can be greatly speeded up by the presence of an inert hydrocarbon or hydrofluorocarbon solvents that dissolve the adsorbate and carry it into the nanomaterial pores.[7] Adsorption capacities are high compared with other more common sorbents such as activated carbon or Ambersorb.

Chemical warfare agents are chemically dismantled (destructively adsorbed) into non-toxic fragments.[6,7] For example, at room temperature paraoxon $[(EtO)_2P(O)OC_6H_4NO_2]$ suffers bond cleavage of all three P-OR bonds as time goes on (as followed by solid state ^{31}P NMR).

Results with intimately mixed oxides show that further enhancement of reactivity for these detoxification reactions can be achieved. Mixed metal oxide systems of $AP-MgO-Al_2O_3$ and $AP-CaO-Al_2O_3$ are better at destructively adsorbing paraoxon than $AP-MgO$, $AP-CaO$ and $AP-Al_2O_3$ by themselves. $AP-MgO-Al_2O_3$ adsorbs all of the (16ul) paraoxon in less than 20 minutes, whereas $AP-Al_2O_3$ takes 60 minutes and $AP-MgO$ adsorbs 15ul in approximately 2 hours. $AP-CaO-Al_2O_3$ also performs well, but not as well as $AP-MgO-Al_2O_3$. Sulfated mixed metal oxides also show further improved adsorption suggesting that increasing the acidity of the sample enhances adsorption.

Acid gases are also efficiently adsorbed at room temperature, and at elevated temperatures the reactions can be driven to solid-gas stoichiometric ratios. For example, a new $SrO-Al_2O_3$ nanomaterial adsorbs H_2S at ambient temperatures. Similarly at 100°C, $AP-CaO-Al_2O_3$ showed good adsorption of

H_2S. The mixed metal oxides out performed AP-SrO at 250°C and 500°C. AP-$CaSrO_2$ gave a molar ratio of 1:1 at 250°C whereas AP-SrO did not perform as well (molar ratio of 1:3). CP-$CaSrO_2$ also gave good results at 500°C, again showing enhanced reactivity over AP-SrO at elevated temperatures. It appears that the mixed metal oxide systems are not as susceptible to crystal growth and reduced activity at higher temperatures, thus making them more suitable for high temperature applications.

At elevated temperatures, chlorocarbon reactions can be driven to stoichiometric proportions, especially if small amounts of transition metal catalysts are added. Thus, CCl_4 reacts stoichiometrically with nano-CaO if a monolayer of Fe_2O_3 is placed on the CaO:[9]

$$2[Fe_2O_3]CaO + CCl_4 \xrightarrow{\ 425°C\ } 2[Fe_2O_3]CaCl_2 + CO_2$$

The catalytic action of Fe_2O_3 (or other transition metal oxides) appears to be due to the transient intermediacy of mobile $FeCl_3$ that then exchanges Cl^- for O^{2-} deep into the CaO nanocrystal.

Catalysis Using Nanoscale Metal Oxides

Examples of nanoscale metal oxides being used as catalysts for dehydrohalogenation, dehydrogenation, and alkylation are also available. Nano-MgO is effective for stripping HCl, HBr, or HI from haloalkanes. This process involves first the conversion of the MgO crystallites to a core/shell structure, for example a $MgCl_2$ coating on remaining MgO, while retaining to an extent the nanostructured form. Then the nanostructured $MgCl_2$ coating is a strong enough Lewis acid to serve as a true catalyst:[10]

$$MgO + BuCl \longrightarrow [MgCl_2]MgO + butene + H_2O$$

$$BuCl \xrightarrow{\ [MgCl_2]MgO\ } butenes + HCl$$

Oxidative dehydrogenation of propane and butane using nano-MgO as a support for vanadium oxide catalysts shows promise.[11] The higher surface areas and unique morphology offered by the MgO serve to enhance catalyst selectivities.

$$propane + O_2 \xrightarrow{\ [VO_x]MgO\ } propene + H_2O$$

The use of iodine cocatalyst/promoter for dehydrogenation of butane has also shown promise.[12]

$$C_4H_{10} + I_2 + MO \rightarrow C_4H_8 + MI_2 + H_2O$$

$$C_4H_8 + I_2 + MO \rightarrow C_4H_6 + MI_2 + H_2O$$

$$2MI_2 + O_2 \rightarrow 2MO + I_2$$

The reaction sequence involves the conversion of a nanomaterial oxide to an iodide followed by conversion back to metal oxide. Here, again, the high surface area of the nano-oxide coupled with its intrinsically higher surface reactivity allows higher activities, selectivities, and lower temperatures.

An additional example of the unusual catalytic properties of nano-MgO is found in the alkylation of toluene or xylene by benzyl chloride.[13] The MgO is a precatalyst that serves as a foundation for the formation of a $MgCl_2$ coating that serves as the true catalyst. In this study it was found that nanocrystal shape was very important, and

$$C_6H_5CH_3 + ClCH_2C_6H_5 \xrightarrow{[MgCl_2]MgO} C_6H_4(CH_3)(CH_2C_6H_5) + HCl$$

hexagonal platelet morphology was more effective than more polyhedral shapes. Apparently the "molecular trafficking" on the surface of the catalyst was more facile when flat surfaces were available.

Biocidal Properties

Nanoscale metal oxides also exhibit biocidal properties due to their abrasive nature, alkaline surfaces, oxidizing power (when elemental halogens are preadsorbed), and the fact that their average particle charge (positive) attracts bacteria (which generally carry overall negative charge).[14] In fact, nano-MgO is biocidal by itself for vegetative cells such as *e-coli* or vegetative *bacillus cereus*. However, for spores, which possess a tough protective outer shell, preadsorbed halogens are necessary. These $MgO \cdot X_2$ halogen adducts are of interest in their own right. The adsorbed halogens (Cl_2, Br_2, ICl, IBr, I_2, ICl_3) are even more chemically reactive than the free halogen molecules. Raman and UV-vis spectra suggest that halogen-halogen bonds are weakened upon adsorption, which apparently is the reason for this enhanced reactivity toward organics and bacteria.[15]

These examples serve to point out the potential usefulness of nanoscale metal oxides in environmental remediation, catalysis, and fine chemical syntheses by "greener" chemistry.

References

1. Klabunde, K.J.; editor, "Nanoscale Materials in Chemistry," Wiley Interscience, New York, 2001, pgs. 1-15.
2. NanoScale Materials, Inc. markets nano-oxides and metals under the tradename NanoActive. Website: www.nanoscalematerialsinc.com.
3. (a) Richards, R. Richards, R.; Li, W.; Decker, S.; Davidson C.; Koper O.; Zaikovski V.; Volodin A.; Rieker T.; Klabunde K.J.; J. Amer. Chem. Soc., 122, 4921-4925 (2000).
   (b) Klabunde, K.J.; Koper, O.; Khaleel, A.; "Porous Pellet Adsorbents Fabricated From Nanocrystals," U.S. Patent 6, 093, 236; July 25, 2000.
4. Kresge, C.T.; Leonowicz, M.E.; Roth, W.J.; Vartuli, J.C.; Beck, J.S.; Nature, 359, 710 (1992).
5. Lamberti, C.; Microporous and Mesoporous Materials, 112, 3589 (1999).
6. (a) Wagner, G.W.; Bartram, P.W.; Koper, O.; Klabunde, K.J.; J. Phys. Chem. B., 103, 3225-3228 (1999).
   (b) Wagner, G.W.; Koper, O.B.; Lucas, E.; Decker, S.; Klabunde, K.J.; J. Phys. Chem. B, 104, 5118-5123 (2000).
   (c) Wagner, G.W.; Procell, L.R.; O'Connor, R.J.; Munavalli, S.; Carnes, C.L.; Kapoor, P.N.; Klabunde, K.J.; J. Am Chem. Soc., 123, 1636-1644 (2001).
7. (a) Rajagopalan, S.; Koper, O.; Decker, S.; Klabunde, K.J.; Chemistry, A European J., 8, 2602-2607 (2002).
   (b) Narske, R.M.; Klabunde, K.J.; Fultz, S.; Langmuir, 18, 4819-4825 (2002).
8. (a) Medine, G.M.; Klabunde, K.J.; Zaikovski, V.; J. Nanoparticle Research, 4, 357-366 (2002).
   (b) Medine, G.; unpublished work.
9. (a) Decker, S.; Klabunde, J.S.; Khaleel, A.; Klabunde, K.J.; Environ. Sci. Tech., 36, 762-768 (2002).
   (b) Decker, S.; Klabunde, K.J.; J. Amer. Chem. Soc., 118, 12465-12466 (1996).
10. Mishakov, I.; Bedilo, A.; Richards, R.; Chesnokov, V.; Volodin, A.; Zaikovskii, V.; Buyanov, R.; Klabunde, K.J.; J. of Catalysis, 206, 40-48 (2002).
11. (a) Pa, K.C.; Bell, A.T.; Tilley, T.D.; J. Catal., 206, 49 (2002).
    (b) Gai, P.L.; Roper, R.; White, M.G.; Solid State Mat. Sci., 6, 401-406 (2002).
12. Chesnekov, V.V.; Bedilo, A.F.; Heroux, D.S.; Mishakov, I.V.; Klabunde, K.J.; J. Catal., in press.
13. Choudary, B.M.; Mulukutla, R.S.; Klabunde, K.J.; J. Amer. Chem. Soc., 125, 2020-2021 (2003).
14. Stoimenov, P.K.; Klinger, R.L.; Marchin, G.L.; Klabunde, K.J.; Langmuir, 18, 6679-6686 (2002).
15. Stoimenov, P.; unpublished work from this laboratory.

Chapter 38

Development of Nanocrystalline Zeolite Materials as Environmental Catalysts

H. Alwy, G. Li, V. H. Grassian, and S. C. Larsen

Department of Chemistry, University of Iowa, Iowa City, IA 52242

Introduction

Nanoscale materials are promising catalysts in the broadly defined field of environmental catalysis. Environmental catalysis refers to the use of catalysts to solve environmental problems, in areas such as waste remediation, emission abatement, and environmentally benign chemical synthesis. Some of the advantages that may be achieved by improved environmental nanocatalyst materials include increased energy efficiency and conversion, reductions in chemical waste and effective waste remediation. The environmental benefits of nanocatalysts include cleaner air and water and, ultimately, a more sustainable future.

Zeolites, which are widely used in applications in separations and catalysis, are aluminosilicate molecular sieves with pores of molecular dimensions. The crystal size of zeolites formed during conventional synthesis range in size from 1,000 to 10,000 nm. Recently, the synthesis of nanometer-sized zeolites has been reported by several groups.(1-12) There has been a great deal of interest in nanocrystalline zeolites because of their properties, such as improved mass transfer and the ability to form zeolite films and other nanostructures. The synthesis and characterization of nanocrystalline NaY zeolites and the formation of transparent thin films from the nanocrystalline NaY are reported here.

Materials and Methods

Nanocrystalline NaY was synthesized using clear solutions according to the method reported in the literature.(1,2) Ludox silica sol (30 wt. %, Aldrich) was deionized with a cation exchange resin, Amberlite (IR-120, Mallinckrodt). 19.7 g of $Al_2(SO_4) \cdot 18H_2O$ (Aldrich, 98%) was dissolved in 75 mL of deionized water. 35 mL of 27% NH_3 solution (Mallinckrodt) was added to precipitate $Al(OH)_3$. After centrifugation, the solid was washed with water and the supernatant was discarded. 54 g of tetramethylammonium hydroxide (TMAOH, 25% aqueous solution, Aldrich) was added to the solid and stirred until a clear solution formed. The clear solution was added to 20 g of deionized Ludox sol. The molar composition of the resulting clear solution was; 2.5 $(TMA)_2O$: $0.041Na_2O$: $1.0 Al_2O_3$: $3.4 SiO_2$: $370 H_2O$. This clear solution was heated in a Teflon-lined stainless steel autoclave for 7 days at 95°C. The resulting solution was centrifuged for 30 min. at 3400 rpm. The product was washed with distilled water and dried in air.

The nanocrystalline NaY product was characterized by powder X-ray diffraction (Siemans D5000) to assess crystallinity and to verify the identity of the zeolite, and by scanning electron microscopy (SEM, Hitachi S-4000), to determine particle size and morphology. Atomic force microscopy (AFM) images were recorded using a Digital Instruments Nanoscope III Scanning Probe Microscope. ^{29}Si (n_L=59.62 MHz) and ^{27}Al (n_L=78.21 MHz) solid-state NMR spectra were obtained using a 300 MHz wide bore magnet with a Tecmag spectrometer with a Bruker 7.5 mm magic angle spinning (MAS) probe with a spinning speed of ~6 kHz. FT-IR spectra were recorded with a Mattson Galaxy 6000 infrared spectrometer equipped with a narrowband MCT detector. Each spectrum was obtained by averaging 500 scans at an instrument resolution of 4 cm^{-1}.

Results and Discussion

Characterization of Nanocrystalline NaY

The XRD pattern (not shown) indicated that NaY zeolite was formed by the hydrothermal synthetic method described in the previous section. SEM images of commercial NaY (Aldrich) and nanocrystalline NaY are shown in Figure 1. For the Aldrich NaY, intergrown crystal agglomerates are observed in the SEM image. For nanocrystalline NaY discrete zeolite crystals are observed with a particle size of 46 ± 8nm. The particle size was obtained by measuring the particle sizes of 50 zeolite crystals in the SEM image of nanocrystalline NaY.

Figure 1. SEM images of Aldrich NaY (left) and nanocrystalline NaY (right).

The samples were further characterized by solid state NMR and FTIR spectroscopy. ^{27}Al and ^{29}Si MAS NMR experiments were performed on the nanocrystalline NaY and the Aldrich NaY for comparison. NMR signals from tetrahedral Al and Si atoms were identified in the ^{27}Al and ^{29}Si NMR spectra, respectively. From ^{29}Si MAS NMR spectra, the Si/Al ratio for nanocrystalline and Aldrich NaY were determined to be 1.7 and 2.4, respectively. By comparing the ^{27}Al MAS NMR spectra (Figure 2) of the nanocrystalline NaY with commercial NaY (Aldrich), an increase in linewidth was observed as the NaY particle size decreases. This line-broadening has been observed before with other zeolites and has been attributed to increased site heterogeneity and increased crystal strain as the particle size decreases.(13) A similar line broadening effect was observed in the ^{29}Si MAS NMR spectra.

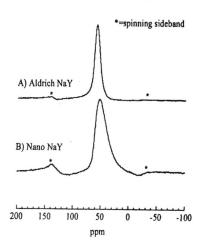

Figure 2. ^{27}Al MAS NMR spectra of A) NaY (Aldrich) and B) nanocrystalline NaY (uncalcined).

Preparation of Transparent High Quality NaY Thin Films

The inherent problem with zeolites as a medium for photooxidation reactions is that zeolites are typically opaque and scatter light. The result is light penetration through the zeolite is small and much of the sample is not exposed to light. For any applications involving intra-cavity photochemistry (e.g. photooxidation)(*14-17*) or zeolite–based optical sensors(*18-20*), transparent zeolite films would be desirable. Films of NaY were prepared by sonication of an aqueous mixture of nancrystalline NaY for several hours. The resulting hydrosol was pipeted onto a pyrex slide and dried in ambient air. Films of commercial NaY (Aldrich) were prepared using the same method. Digital images of the NaY films are shown in Figure 3. In each case, the film was prepared using approximately the same mass of NaY zeolite. The film prepared from the nanocrystalline NaY hydrosol is much more uniform than the film prepared from the Aldrich NaY hydrosol. The increased transparency of the films can be observed visually. The "Y" printed on the paper behind the film can be clearly seen through the nanocrystalline NaY film (right) but is much more difficult to see through the Aldrich NaY film (left). To obtain more quantitative information, the percent transmittance was measured using UV/Vis spectroscopy. The nanocrystalline NaY film had a percent transmittance of 70-80% in the 300-700 nm range compared to a precent transmittance of 30-40% for the Aldrich NaY film in the same range.

Figure 3. Films prepared from hydrosols of Aldrich NaY (left) and nanocrystalline NaY (right).

Images of the nanocrystalline NaY film were obtained using atomic force microscopy (AFM) are shown in Figure 4. The AFM images of the zeolite films show that the film surface is continuous and smooth on the hundreds of nanometer length scale. Attempts to obtain AFM images of the commercial Aldrich NaY films were unsuccessful due to the extreme roughness of the film.

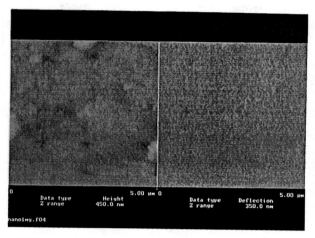

Figure 4. AFM images of the nanocrystalline NaY film obtained in the height mode (left) and deflection mode (right).

Spectroscopic Investigations of NO₂ Adsorbed on Nanocrystalline NaY

FTIR spectroscopy was also used to characterize the nanocrystalline NaY and to monitor the adsorption, desorption and reaction of various molecules in the zeolite. The FTIR spectra of the adsorption of NO_2 at progressively increasing partial pressures of NO_2 on nanocrystalline NaY are shown in Figure 5. Spectral features observed can be identified as NO^+ (2000-2180 cm^{-1} spectral region) and NO_3^- (1300-1600 cm^{-1} spectral region). These species are proposed to form through a cooperative effect whereby two adsorbed NO_2 molecules in close proximity to one another autoionize according to the reaction:

$$(NO_2)_2 \rightarrow NO^+ + NO_3^-$$

Further FTIR studies of the nanocrystalline NaY are in progress to examine the reactivity of the nanocrystalline NaY for NO_x decomposition reactions.

In conclusion, nanocrystalline NaY was successfully synthesized with a 46 ± 8 nm particle size. The NaY was extensively characterized using a variety of techniques including XRD, SEM, AFM, UV/Vis, solid state MAS NMR and FTIR spectroscopy. Transparent, thin films were prepared from the nanocrystalline NaY. Future studies will focus on applications of the nanocrystalline NaY as environmental catalysts.

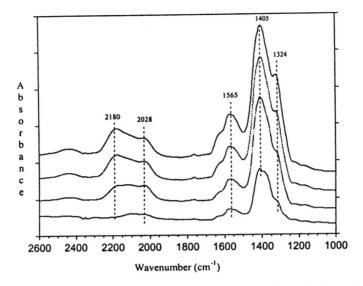

Figure 5. FTIR spectra of NO₂ adsorption on nanocrystalline NaY as a function of increasing NO₂ pressure (100-1000 mTorr from bottom to top

Acknowledgements

Dr. D. Stec and C. Jones are acknowledged for assistance with NMR experiments and B. J. Kruegger is acknowledged for assistance with AFM experiments. The research described in this article has been funded by the Environmental Protection Agency through EPA grant no: R82960001 to SCL and VHG.

References

(1) Castagnola, N. B.; Dutta, P. K. *J. Phys. Chem. B* **2001**, *105*, 1537-1542.

(2) Schoeman, B. J.; Sterte, J.; Otterstedt, J.-E. *Zeolites* **1994**, *14*, 110-116.

(3) Li, Q.; Creaser, D.; Sterte, J. *Chem. Mater.* **2002**, *14*, 1319-1324.

(4) Zhu, G.; Qiu, S.; Yu, J.; Sakamoto, Y.; Xiao, F.; Xu, R.; Terasaki, O. *Chem. Mater.* **1998**, *10*, 1483-1486.

(5) Camblor, M. A.; Corma, A.; Mifsud, A.; Perez-Pariente, J.; Valencia, S. *In Progress in Zeolites and Microporous Materials*, 1997; Vol. 105, pp 341-348.

(6) Holmberg, B. A.; Wang, H.; Norbeck, J. M.; Yan, Y. *Microp. and Mesop. Mater.* **2003**, *59*, 13-28.

(7) Jacobsen, C. J. H.; Madsen, C.; Janssens, T. V. W.; Jakobsen, H. J.; Skibsted, J. *Microporous Mesoporous Mater.* **2000**, *39*, 393-401.

(8) Jung, K. T.; Shul, Y. G. *Chem. Mater.* **1997**, *9*, 420-422.

(9) Mintova, S.; Olsen, N. H.; Valtchev, V.; Bein, T. *Science* **1999**, *283*, 958-960.

(10) Mintova, S.; Valtchev, V. *Microp. and Mesop. Mater.* **2002**, *55*, 171-179.

(11) Reding, G.; Maurer, T.; Kraushaar-Czarnetzki, B. *Micropor. Mesopor. Mater.* **2003**, *57*, 83-92.

(12) Tsapatsis, M.; Lovallo, M.; Okubo, T.; Davis, M. E.; Sadakata, M. *Chem. Mater.* **1995**, *7*, 1734- 1741.

(13) Zhang, W.; Bao, X.; Wang, X. *Catalysis Letters* **1999**, *60*, 89-94.

(14) Pitchumani, K.; Joy, A.; Prevost, N.; Ramamurthy, V. *Chem. Commun.* **1997**, 127-130.

(15) Robbins, R. J.; Ramamurthy, V. *Chem. Commun.* **1997**, 1071-1072.

(16) Xiang, Y.; Larsen, S. C.; Grassian, V. H. *J. Am. Chem. Soc.* **1999**, *121*, 5063-5072.

(17) Li, X.; Ramamurthy, V. *J. Am. Chem. Soc.* **1996**, *118*, 10666-10667.

(18) Remillard, J.; Jones, J.; Poindexter, B.; Narula, C.; Weber, W. *Applied Optics* **1999**, *38*, 5306-5309.

(19) Meinershagen, J. L.; Bein, T. *J. Am. Chem. Soc.* **1999**, *121*, 448-449.

(20) Mintova, S.; Mo, S.; Bein, T. *Chem. mater.* **2001**, *13*, 901-905.

Chapter 39

A Vibrational Spectroscopic Study of the CO + NO Reaction: From Pd Single Crystals at Ultrahigh Vacuum to Pd Clusters Supported on SiO$_2$ Thin Films at Elevated Pressures

Emrah Ozensoy[1], Christian Hess[1,2], Ashok K. Santra[1],
Byoung Koun Min[1], and D. Wayne Goodman[1,*]

[1]Department of Chemistry, Texas A&M University, College
Station, TX 77842
[2]Current address: Department of Chemistry, University of California,
Berkeley, CA 94720
*Corresponding author: goodman@mail.chem.tamu.edu

1. Introduction

Design and development of three-way catalytic converters have received significant scientific attention in the catalysis field due to their commercial importance in automobile exhaust emissions (for the simultaneous reduction of NO$_x$ and oxidation of CO and unburned hydrocarbons). Although the reaction between CO and NO has been studied over various transition and noble metal catalysts, only recently Pd-based catalysts have been proposed as an alternative to the Pt/Rh catalysts that are predominantly used.(1) Hence, a fundamental understanding of the reaction pathways of the CO + NO reaction on Pd surfaces is of vital importance. Adsorption of CO and NO on low index planes of palladium has been studied in detail under ultra-high vacuum conditions(1), however only a few studies have examined the CO + NO reaction in real time at elevated pressures.(4-8)   These studies on a Pd(111) surface (4-8) have

284   © 2005 American Chemical Society

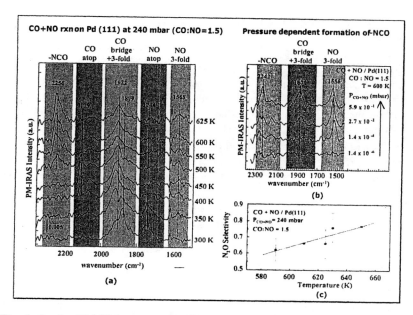

Fig. 1. *In situ PM-IRA spectra for CO+NO reaction on Pd(111) and diagram showing the temperature dependence of N₂O selectivity of the model catalyst. All the spectra are obtained in the presence of CO + NO gas phase (P$_{CO}$/P$_{NO}$ = 1.5) at the given catalyst temperatures.*

focused on the reaction kinetics and the branching ratio between the desired reaction pathway (1) and the less ideal pathway (2).

$$CO + NO \rightarrow CO_2 + \tfrac{1}{2} N_2 \tag{1}$$
$$CO + 2 NO \rightarrow CO_2 + N_2O. \tag{2}$$

Besides the gas phase products, adsorbed isocyanate (-NCO) is one of the potentially important products of the CO + NO reaction.[7-8] There have been theoretical[9] and experimental[10-12] studies of isocyanate on various well-defined transition metal surfaces; however, metal-bound isocyanate species on well-defined Pd catalysts have not been reported yet. Therefore, the role of such species in the CO + NO reaction is still unexplored. In this study, we report the formation of isocyanate in the CO + NO reaction on Pd(111) and discuss its relevance to the kinetics and mechanism of the CO + NO reaction on Pd using polarization modulation infrared reflection absorption spectroscopy (PM-IRAS).

Although studies on well-defined singe crystal model catalysts provide valuable information about heterogeneous catalytic reactions, some of the profoundly important aspects of the real working catalysts (consisting of metal particles dispersed on an oxide surface) such as metal cluster-support interaction, sintering, morphology, size, and cluster restructuring under catalytic conditions cannot be addressed in single crystal model catalyst studies. Therefore we have also synthesized and studied a more complex model catalyst consisting of Pd nano-clusters supported by crystalline SiO_2 grown epitaxially on a Mo(112) substrate. Furthermore, by combining scanning tunneling microscopy (STM) and *in situ* PM- IRAS we have investigated the activity of supported Pd nano-clusters toward CO dissociation.

2. Materials and Methods

The PM-IRAS data were acquired in a combined ultrahigh vacuum (UHV) - micro-reactor system equipped with Auger electron spectroscopy (AES), temperature programmed desorption (TPD) and low energy electron diffraction (LEED). The STM experiments were carried out in a separate UHV chamber equipped with STM, X-ray photoelectron spectroscopy (XPS), LEED, and AES. Typically the STM images were acquired in UHV in the constant current mode with a 1.5 - 2.5 V tip bias and ~0.1nA tunneling current. For further experimental details, the reader is referred to our previous publications.*(6-8, 13)*

3. Results and Discussion

In the first part of our studies, we have investigated the CO + NO reaction on Pd(111) at pressures up to 240 mbar using *in situ* PM-IRAS (see fig.1), a technique that simultaneously provides information on the *in situ* adsorption behavior of the surface species as well as the reaction kinetics via gas-phase infrared absorption. Therefore the species on the catalyst surface under reaction conditions can be examined and correlated with kinetic parameters such as conversion and selectivity.

The formation of an isocyanate (-NCO) species on Pd(111) under reaction conditions (T $\geq$ 500K), is observed by PM-IRAS (fig. 1). The -NCO species is found to be stable within 300 - 625 K, i.e., even outside the reaction regime of

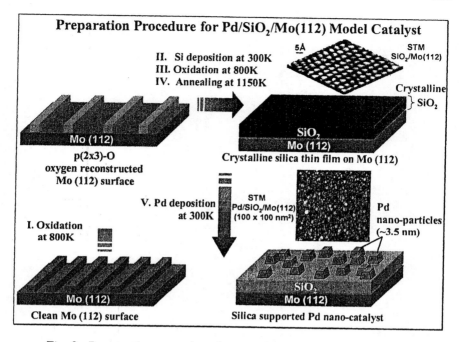

Fig. 2. Preparation procedure for a model catalyst consisting of Pd nano-clusters supported by a crystalline SiO₂ ultra-thin film.

the CO + NO reaction. The observation of –NCO provides valuable information regarding the mechanism of the CO + NO reaction since its formation implies the dissociation of NO, a crucial step in the CO + NO reaction. It has been shown by gas phase absorption infrared spectroscopy that the measured activities for CO_2 and N_2O production, N_2O selectivities and the apparent activation energy based on the CO_2 yields (54 ± 21 kJ/mol) are in good agreement with previous results obtained at lower pressures.*(4)* At high temperatures, the higher $[NO]_{ads}/[CO]_{ads}$ ratio leads to the less desirable reaction pathway (2) being dominant since it facilitates a higher conversion of adsorbed NO. Considering the stability of the isocyanate species and the similar reactivity behavior of the CO + NO reaction in high-pressure and in low-pressure experiments, each of which show the formation of isocyanate, it is likely that under the conditions studied here isocyanate plays the role of a spectator rather than an intermediate in the CO + NO reaction. However, the fact that the formation of isocyanate can only be observed at elevated pressures shows that *in situ* studies under more "realistic" conditions is necessary to obtain a detailed understanding at least for certain surface reactions.

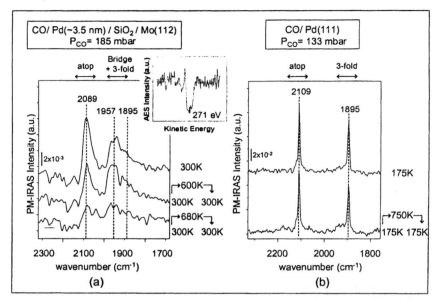

Fig. 3. Reversibility/irreversibility of CO adsorption on silica supported Pd nano-clusters and Pd (111) single crystal surface with temperature. Inset in fig. 3a presents the difference between AES for Pd nano-clusters before and after CO adsorption at P_{CO} = 185 mbar and T = 680K. The feature at 271 eV corresponds to carbon.

In the second part of our studies, STM and *in situ* PM-IRAS are combined to study CO adsorption on a complex model catalyst at elevated pressures (P_{CO} = 185 mbar) and temperatures (300 - 680 K). The model catalyst consists of Pd clusters deposited on a crystalline SiO_2 ultra-thin film support grown on a Mo(112) substrate (fig. 2). These results are compared with our previous *in situ* PM-IRAS results[5] for CO adsorption on a Pd (111) single crystal at elevated pressures (10^{-6} - 600 mbar) and temperatures (220 and 750 K) as shown in fig. 3. STM and *in situ* PM-IRAS data for CO adsorption on Pd (111) and silica supported Pd nano-clusters indicate that Pd nano-clusters exhibit primarily <111> facets with a small contribution from the <100> facets and have an average height and diameter of 0.7 and 3.5 nm, respectively. Furthermore, compared to atomically flat Pd(111), Pd nano-clusters have a relatively high curvature and a relatively high density of surface defects such as steps or kinks that facilitate CO adsorption onto atop sites rather than multiply coordinated sites. In addition our *in situ* PM-IRAS results reveal that CO dissociation and subsequent surface poisoning takes place on the silica supported Pd nano-clusters at elevated pressures (P_{CO} = 185 mbar) and temperatures (T > 600 K). In contrast to Pd nanoparticles, CO adsorb molecularly and reversibly on the flat Pd(111) surface without dissociation. The activity of the Pd nano-structures for CO dissociation is attributed to defect sites on the nano-clusters that exist as a result of their three-dimensional morphology.

Acknowledgements

We acknowledge with pleasure the support of this work by the Department of Energy, Office of Basic Energy Sciences, Division of Chemical Sciences.

References

1. Heck, R. M.; Farrauto, R. J. 1995, *Catalytic Air Pollution Control: Commercial Technology,* International Thomson Publishing: New York, NY.
2. Hoffmann, F. M., *Surf. Sci. Rep.* **1983**, 3, 107
3. Brown, W. A.; King, D. A, *J. Phys. Chem. B.* **2000**, *104*, 2578.
4. S.M. Vesecky, P.J. Chen, X. Xu and D.W. Goodman, *J. Vac. Sci. Technol. A* **1995**, *13*, 1539.
5. Oh, S.H.; Fisher, G.B.; Carpenter, J.E.; Goodman, D.W.; *J. Catal.* **1986**, *100*, 360.
6. Ozensoy, E.; Meier, D. C.; Goodman, D. W., *J. Phys. Chem.* **2000**, *106*, 9367 and references therein.
7. Ozensoy, E.; Hess, Ch.; Goodman, D. W., *J. Am. Chem. Soc.* **2002**, *124*, 8524.
8. Hess, Ch.; Ozensoy, E.; Goodman, D. W., *J. Phys. Chem. B.* **2003**, *107*, 2759.
9. Garda, G.R.; Ferullo, R.M.; Castellani, N.J., *Surf. Rev. Lett.* **2002**, *8*, 641.
10. Celio, H.; Mudalige, K.; Mills, P. Trenary, M., *Surf. Sci.* **1997**, *394*, L168.
11. Yang, H.; Whitten, J. L., *Surf. Sci.* **1998**, *401*, 312.
12. Miners J. H.; Bradshaw A.M.; Gardner P., *Phys. Chem. Chem. Phys.* **1999**, *20*, 4909.
13. Ozensoy E.; Min, B. K. Goodman, D. W. to be published.

Chapter 40

Nanostructured Catalysts for Environmental Applications

S. Ismat Shah[1,2], A. Rumaiz[2], and W. Li[1]

Departments of [1]Materials Science and Engineering and [2]Physics and Astronomy, University of Delaware, Newark, DE 19716

Introduction

There are two areas of interest in environmental technology where nanocatalysts are potentially useful. These areas include (i) reduction of the pollutant emissions (ii) decrease of the concentration of already existing pollutants. Two specific materials to exemplify these two types of nanocatalysts are discussed. TiO_2 has been known to be a photocatalyst for the degradation of various organic and inorganic pollutants.[1-3] The nanostructure plays in increasing the photocatalytic efficiency of TiO_2. Doping with the appropriate dopant can also help increase the photocatalytic efficiency and cause red shift in the band gap of TiO_2 making it absorb in the visible range. For emission control of the pollutants, specifically, for NO_x, a precious metal (Pt, Pd, Ru, Rh etc.) or a toxic gas (ammonia) is used. There is a need for an alternative catalyst. A potentially important and low cost alternative is WC_x.[4-5] The synthesis and characterization of WC_x nanoparticles is described.

Method

Doped and undoped TiO_2 nanoparticles were synthesized and investigated for the purpose of enhancement of photoreactivity. Due to the importance of the anatase structure of TiO_2 in the photocatalysis application, the nanoparticle size effect on the phase transformation of pure anatase to rutile was studied. Based on our previous work,[1-3] lanthanide Nd^{3+} ion was selected to improve the photoreactivity of TiO_2, especially, the visible light photoactivity.

All samples were prepared by metalorganic chemical vapor deposition (MOCVD) which was described in our previous studies.[1-2] $Ti[OCH(CH_3)_2]_4$ (titanium tetraisopropoxide, TTIP, Aldrich 97%) was used as Ti precursor. 99.999% Ar (3 Torr, 3 sccm) and 99.999% O_2 (10 Torr. and $20-35$ sccm) were used as carrier gas and reactant gas, respectively. The total deposition pressure was about 16 Torr. 3.5 cm diameter discs made of multiple layers of 400×400-mesh stainless steel screens were used as substrates and held perpendicular to the flow direction of the gases in the reaction chamber, to collect the particles. The bath temperature of TTIP and deposition temperature were 220 and 600 °C, respectively. $[CH_3COCHCOCH_3]_3Nd \cdot xH_2O$ (neodymium (III) acetylacetonate hydrate, Aldrich) was used as Nd^{3+} dopant precursor. The melting point of dopant precursor is 143 °C. The precursor, in the powder form, was placed in a ceramic container which was directly placed in a position of the low temperature inlet region of the reactor. The position of the container was calibrated for temperature and, therefore, the effusion rate of the precursor to obtain the desired dopant concentration.

Results and Discussion

Nanocrystalline TiO_2 samples with different particles sizes, 13, 19, and 25 nm, were used to investigate the thermal properties by X-ray Diffraction (XRD) and Transmission Electron Spectroscopy (TEM). Each sample was divided into several portions for different temperature (700, 750, 800°C for 1 hour) annealing. Results showed that the enthalpy of transformation of anatase into rutile was dependent on the particles size. The percentage of each phase was measured from XRD by using formula: $W_R = A_R/A_0 = A_R/(0.884A_A+A_R)$,[6] where, W_R is rutile percent, $A_0 = 0.884A_A+A_R$, A_A and A_R are integrated intensities of anatase and rutile (101) and (110) peaks, respectively. The XRD results showed that the onset of phase transform varied with the particle size. Also, 25 nm size particles show 0.015 wt.% rutile phase whereas 13 nm particles show 0.093 wt.% of rutile precipitated at 700°C. The activation energy of the transformation was calculated based on the XRD results. . The slopes of two linear fits are different, 298.85 and 180.28 kJ/mol, for 25 and 13 nm particles, respectively. Due to the larger surface area of finer particles, its contribution to the total energy of the particle is higher. There is also larger contribution from the surface stress energy.[7] Thus, if more total free energy of exists in the

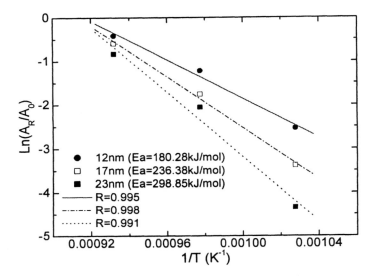

*Figure 1 is the arrenhius plot of ln(A_R/A_0) vs. 1/T which was
used for the activation energy calculations*

smaller anatase particles, the difference of total Gibbs free energies for rutile
and anatase will be more negative, which is the driving force of anatase to rutile
transformation. Hence, the 13 nm particles show lower activation energy and
lower anatase – rutile transformation onset temperature

The effect of doping on the optical and photocatalytic properties of TiO_2
nanoparticles was studied. Nanoparticles with Nd^{3+} concentrations ranging from
0 to 1.5 at.% were prepared. Nd^{3+} concentrations were determined by XPS and
EDS. The size of particles was 21-25 nm and all samples had anatase phase
with no separate dopant related phase, as analyzed by XRD and TEM. The
samples were dispersed in methanol for UV-VIS light absorption experiments.
These experiments showed red shift of absorption edge for all doped samples
and the sample with 1.5 at% doping had the largest absorption over a range of
470 nm. The results are plotted in Figure 2.

Nanostructured WC_x was reactively sputtered in argon- methane plasma.
Samples were collected on glass, quartz, silicon and sapphire substrates at
temperatures up to 700 °C. The carbon content in the samples was varied by
changing the partial pressure of methane. X-ray diffraction (XRD) analysis was
carried out to analyze the crystalline structure. Figure 3 shows the XRD pattern
of a sample grown on Si substrate at 700 °C. The (400) and the (200) diffraction
peaks corresponds to the Si substrate. All the other peaks correspond to W_2C.
Samples grown at room temperature were amorphous.

The carbon content in the film was analyzed using X-ray Photon
Spectroscopy (XPS) and Rutherford Backscattering Spectroscopy (RBS).

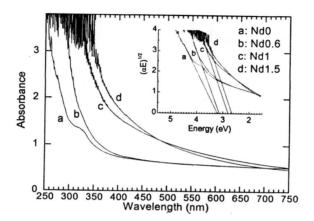

Fig. 2. Light absorption and band gap calculations

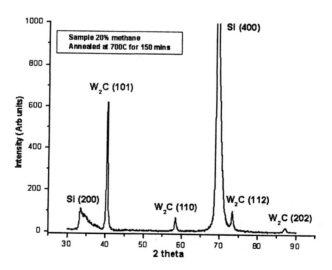

Fig. 3. XRD pattern of WC$_x$ nanoparticles.

Figure 4 shows the XPS spectra of W 4f region for a sample deposited with 20% CH_4. Pure W has a peak at BE of 31.0 while WO_x and WC_x has peaks between 32.5-37.0 eV and 31.4-31.8 eV, respectively. The XPS spectrum was fitted with four peaks and shows the presence of WC_x and WO_x phases. However, RBS done using a 2 MeV He^+ beam (National Electronics Corporation 5 SDH2) and the RUMP simulation (5) show no oxygen in the particles. Therefore, the oxygen related phase is present only at the surface due to atmospheric exposure.

To analyze the grain size films were sputtered on copper mesh and studied under TEM (JEM2010F FasTEM). To examine the effect of temperature on grain size films were sputtered on mesh at temperature ranging from room temperature to 600 °C. The average grain size was found to be around 1.5-2.0 nm. The grain size also remained constant for temperature up to 600 °C. Figure 5 shows the typical TEM micrograph.

The optical absorption results suggest that for large band gap semiconductor TiO_2 nanoparticles, visible light absorption is possible by Nd^{3+} doping. The experimental results are consistent with the theoretical calculation of density of electron states (DOS) obtained by using a linearized augmented plane wave simulation.[8] The visible light absorption originates from the band gap tailoring by Nd doping and is confirmed by near edge X-ray absorption fine structure (NEXAFS) measurements. By plotting $(\alpha E)^{1/2}$ vs. E (α is absorption coefficient), the band gap energies can be calculated from light absorption, which are 3.22, 3.12, 2.81, and 2.67 eV for 0, 0.6, 1, 1.5 at% Nd doping levels, respectively. These data are also consistent with NEXAFS measurements.[3] Nd^{3+} ions have $4f^3$ characteristic electrons, especially for substitutional Nd^{3+} in Ti^{4+} positions, strong electronic interactions occur with the conduction band

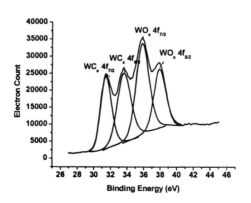

Fig. 4. XPS of W 4f region.

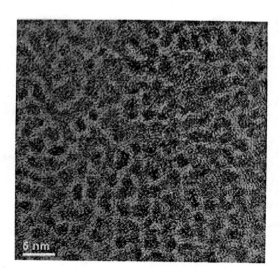

Fig. 5. TEM of WC$_x$ nanoparticles.

electrons of TiO$_2$. Hence, Nd *4f* level forms new lowest unoccupied level in TiO$_2$ and the electron excitation from O *2p*, the highest occupied level, to Nd *4f* orbital becomes possible. Therefore, TiO$_2$ band gap changes with the incorporation of Nd^{3+} ions. The photodegradation experiments of 2-chlorphenol solutions under UV irradiation exhibited high reaction rate with Nd^{3+} doped sample.[1] Visible light photoreactivity of Nd^{3+} doped TiO$_2$ nanoparticles are under investigation.

In the synthesis of WC$_x$ by reactive sputtering, target poisoning behavior is an issue that causes the target current and sputtering rate of the target to go down as the partial pressure of the reactive gas (CH$_4$) increases above a critical limit. The critical methane concentration at which the metal – poison mode transition occurs was measured to be around 40% of CH$_4$ in Ar.

Some of the issues in reactive sputtering can be overcome by using a unique hollow cathode source (HCS). In this method the target sheet is fixed on a tubular water cooled copper tube which acts like a cathode. In this method the sputtered flux is uniform inside the cathode and the plasma density is also high compared with planar magnetron. Experiments done in this arrangement showed no hysteresis behavior in target current. The particles form within the hollow cathode is purges out by the flowing gas. The nanoparticle size is a function of the pressure and the target power.

References

1. W. Li, S. Ismat Shah, C.-P. Huang, O. Juang, and C. Ni, (2002) Mater. Sci. Eng. B **96**, 247.
2. W. Li, S. Ismat Shah, M. Sung, and C.-P. Huang, (2002) J. Vac. Sci. Technol. B **20**, 2303.
3. A. Burns, W. Li, C. Baker, and S. Ismat Shah, (2002) Mater. Res. Soc. Sym. Proc. **703**, 193.
4. P.N. Ross, P. Stonehart, (1975) J. Catal. **39**, 298.
5. M. Zhang, H.H. Hwu, M.T. Buelow, J.G. Chen, T.H. Ballinger, P.J. Anderson, (2001) Catal. Letters **77**, 29.
6. A. A. Gribb and J. F. Banfield, (1997) Am. Mineral. **82**, 717.
7. H. Zhang and J. F. Banfield, J. Mater. Chem.(1998) **8**, 2073.
8. W. Li, S. Ismat Shah, D. Doren and Y. Wang, unpublished

Chapter 41

Catalytic Oxidations Using Nanosized Octahedral Molecular Sieves

Young-Chan Son[1], Jia Liu[2], Ruma Ghosh[1], Vinit D. Makwana[1], and Steven L. Suib[1-3,*]

[1]Department of Chemistry, University of Connecticut, U–3060 55 North Eagleville Road, Storrs, CT 06269–3060
[2]Institute of Materials Science, University of Connecticut, Storrs, CT 06269
[3]Department of Chemical Engineering, University of Connecticut, Storrs, CT 06269
*Corresponding author: suib@uconn.edu

Introduction and Background

Considerable efforts have been invested to accomplish hydrocarbon oxidation using heterogeneous catalysts with excellent properties [1-4]. Readily available starting chemicals add valuable versatile chemicals are the main targets for the petroleum industry. Oxidations are widely accepted procedures for synthesizing intermediates for the bulk chemical and pharmaceutical industry. However, this reaction is hard to perform and requires extreme conditions. High pressures and high temperatures are usually required. Gas phase reactions are non-selective, and carbon monoxide and carbon dioxide are formed due to complete oxidation. Environmentally friendly reactions of oxidation have received much attention by industry. Cost efficient processes such as solvent free, low energy consumption, and non-pressurized methods are desirable. The development of a catalytic, highly selective, energy efficient, and environmentally friendly process of oxidation is highly desirable. Stoichiometric oxidations of chemical compounds by active manganese oxides have been known for a long time. Stoichiometric oxidation is unattractive and catalytic oxidations will replace traditional stoichiometric methods.

The synthetic OMS (Octahedral Molecular Sieves) materials, which occur as cryptomelane in nature, have been reported for alcohol oxidations [5]. Alcohols can be easily converted to aldehydes and ketones. The kinetics and mechanism of alcohol oxidation have also been also reported [6]. Based on these reports, alcohol oxidations can be broadened to hydrocarbon oxidations. OMS materials can be utilized for the oxidation of hydrocarbons. Various stoichiometric oxidations of hydrocarbons were reported. Potassium permanganate adsorbed on a solid support oxidized alkylbenzene at the benzylic position [7]. These oxidations mostly require stoichiometric oxidants. Strong acids, halogen compounds, and halogenated solvents are usually required.

Material and Methods

OMS has been synthesized and applied for a variety of catalytic applications [8]. OMS has a 2 × 2 tunnel structure and dimensions of 4.6 Å × 4.6 Å (Figure 1). The OMS-2 has variable oxidation states of Mn^{2+}, Mn^{3+}, and Mn^{4+} [9]. The average oxidation state of OMS-2 is 3.8. The mixed valence of manganese is known to be a critical factor, which is responsible for carrying out the oxidation. OMS-2 materials are known to be microporous materials, which act like zeolite-like materials. Methods of synthesizing well-defined OMS-2 materials are reported and characterized [9,10]. OMS synthesis is easy and inexpensive. By inserting divalent cations into OMS, electronic, catalytic, and structural properties can be varied. Nano-sized OMS materials have been synthesized using cross-linking reagents (PVA, glycerol and glucose) [11]. High resolution SEM (scanning electron microscopy) shows that these nano-sized OMS materials are nanofibers or nanorods.

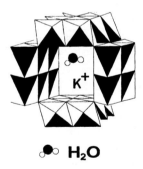

$\bullet$ H_2O

Figure 1. Structure of K-OMS-2

Liquid phase oxidations of toluene or α,β unsaturated ketones with or without peroxide have been carried out depending on substrate. When peroxide (mostly TBHP) was not used, oxygen was the oxidant for the reaction. The oxidation of toluene with molecular oxygen in the liquid phase is carried out by radical initiators. Propagation and termination are major factors to determine the reaction rate and product distribution. The presence of radicals has not been confirmed for OMS system.

Results and Discussion

OMS has Lewis and Brönsted acid and base sites which help the oxidation process, depending on reaction conditions. The characteristics of porosity of OMS are also important parameters for these reactions. Toluene and α,β unsaturated ketones were selected as substrates. Toluene was converted to benzyl alcohol, benzaldehyde, and benzoic acid using AIBN. Oxidation of toluene under reflux conditions using nano-sized OMS materials gave 10 ~ 15 % conversion without any solvent. Selectivity was changed depending on catalyst, reaction temperature, and the amount of catalyst. α,β unsaturated ketones gave 1,4 diketone product using TBHP. Using TBHP as an oxidant, more valuable chemicals were formed. Fluorene was exclusively transformed to 9-fluorenone using OMS and TBHP. α- isophorone was oxidized to 4-ketoisophorone using nano-sized OMS and TBHP in 30 % conversion and 98 % selectivity.

Different kinds of OMS catalysts gave various products. The selectivity was excellent for certain conditions. Nanosized OMS gave higher activity. Conventional OMS materials, which are made by the reflux method, provide higher selectivity but less activity.

$$R\!-\!H \; + \; A\cdot \longrightarrow R\cdot \; + \; AH$$

$$R\cdot \; + \; O_2 \longrightarrow ROO\cdot$$

$$2ROO\cdot \longrightarrow 2RO\cdot \; + \; O_2$$

$$RO\cdot \; + \; R\!-\!H \longrightarrow ROH \; + \; R\cdot$$

R: Alkyl or Benzyl
A: Radical Initiator

Figure 2. General free radical mechanism of hydrocarbon oxidation

Oxidation of aromatic hydrocarbons may be due to the propagation of mixed free radicals as shown in the above example. The mixed valence of manganese species and oxygen also plays an important role (Figure 3). Oxidation with TBHP as well as OMS materials also gave very promising results. OMS catalyst is the key, which gave oxygen-containing product. High selectivity may be due to the porosity of the OMS catalysts.

300

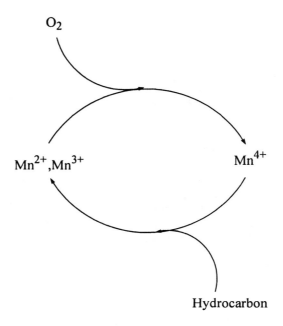

Figure 3. General mechanism of hydrocarbon air oxidation using OMS

This preliminary study gave a fundamental view of the catalytic oxidation using nano-sized OMS materials. The conversions of easily available raw materials to industrially useful chemicals using environmentally friendly nano-sized catalysts are very important.

References

1. Sheldon, R.A., Kochi, J.K. (1981). *Metal-Catalyzed Oxidations of Organic Compounds:* Academic Press: New York.
2. Shilov, A.E., Shul'pin, G.B. (1997). *Chem. Rev.* **97**, 2879-2932.
3. Yu, J.Q., Corey, E.J. (2003). *J. Am. Chem. Soc.* **125**, 3232-3233.
4. Parvulescu, V., Anastasescu, C., Constantin, C., Su, B.L. (2003). *Catal. Today* **78**, 477-485.
5. Son, Y.C., Makwana, V.D., Howell, A.R. and Suib, S.L. (2001). *Angew. Chem. Int. Ed.* **40**, 4280-4283.
6. Makwana, V.D., Son, Y.C., Howell, A.R. and Suib, S.L. (2002). *J. Catal.* **210**, 46-52.
7. Noureldin, N.A., Zhao, D., Lee, D.G. (1997). *J. Org. Chem.* **62**, 8767-8772.

8. Brock, S.L., Duan, N., Tian, Z.R., Giraldo, O., Zhou, H., Suib, S.L. (1998). *Chem. Mater.* **10**, 2619-2628.
9. DeGuzman, R.N., Shen, Y.F., Neth, E.J., Suib, S.L., O'Young, C.L., Levine, S., Newsam, J.M. (1994). *Chem. Mater.* **6**, 815-821.
10. Yin, Y.G., Xu, W.Q., DeGuzman, R., Suib, S.L., O'Young, C.L. (1994). *Inorg. Chem.* **33**, 4384-4389.
11. Liu, J., Cai, J., Shen, X., Suib, S. L., Aindow, M., (2003). *MRS sym. Proc.* **755**, DD6.24.

Environmental Applications: Environmentally Benign Nanomanufacturing

Editors
Ajay P. Malshe
K. P. Rajurkar

Chapter 42

Environmental Applications: Environmentally Benign Nanomanufacturing

An Overview of Nanomanufacturing and Progress

Ajay P. Malshe[1] and K. P. Rajurkar[2]

[1]SERC for Durable Micro and Nano Systems, Department of Mechanical
Engineering, University of Arkansas, Fayetteville, AR 72701
(telephone: 479–575–6561; email: apm2@engr.uark.edu)
[2]Center for Non Traditional Manufacturing Research, Department of
Industrial and Management Systems Engineering, University of Nebraska,
Lincoln, NE 68588

Introduction

From SMART tools to MEMS and NEMS, SMART cards to cell phones,
micro/nano satellites to space system-on-chip, quantum computing to DNA
computing, research on nano and micro systems engineering is leading to
changes in the way we live. Such miniaturized systems are desired for various
applications as components in automobiles, aerospace vehicles, bio-medicine,
informatics hardware, high performance computing, electronics, etc. Further, the
events of September 11th escalated the demand for the next generation
engineered systems for on and off homeland security, particularly in identifying
the signs before such unfortunate events occur. The research and progress in
nano-manufacturing harbors on the platform of micro and meso size tools and
subsystems, which are then interfaced to the macroscopic world. Thus the
research in nano as well as micro/meso materials, processes and systems enables
important conventional requirements such as efficiency, portability, robustness,
programmability, cost efficiency, durability, and new set of requirements such
as configurability, evolutionary and adaptability, intelligent decision making
abilities, realtime interacting and self powered.

The last century (20th Century) has observed inception and impressive
growth of microelectronics technology and related IC based systems. But it is

important to remember that the challenges of the present century (21st) are not only to continue to enjoy the benefits of microelectronics and IC technology but also to explore the avenues for manufacturing multi-signal (electrical, optical, chemical, biological, etc.) / multifunctional *integrated* systems at nano and micro scales. One of the major trends in nano integrated micro technology is the attraction and desire to integrate nano/micro - bio - info . It is imperative that the future of science and engineering of nanomanufacturing relies significantly on the development of fundamental understanding of materials, processes and sub-system and system behavior as well as the ability to design, synthesize, process, measure and manufacture reproducible sub-systems and systems at the nano scale.

Following text provides detailed insight into the subject gathered during NSF-EC Workshop on Nanomanufacturing and Processing held in San Juan, Puerto Rico on 5-7 January 2002. The Workshop was held within the framework of the cooperation between the National Science Foundation (NSF) and the European Commission (EC, materials sciences), with the aim of catalyzing progress in research and education in the emerging field of Nanomanufacturing and processing.

Nanomanufacturing: Definition And Promise

Nanomanufacturing encompasses all processes aimed toward building of nanoscale (in 1D, 2D, or 3D) structures, features, devices, and systems suitable for integration across higher dimensional scales (micro-, meso- and macroscale) to provide functional products and useful services. Nanomanufacturing includes both bottom-up and top-down processes. While clearly not comprehensive, Table 1 gives examples of a few nanomanufacturing processes.

With both bottom-up and top-down methods, materials can be manipulated at the nanoscale to obtain specified positioning, size, and/or structure. The question that arises, however, is how to scale up these processes to robust, high volume, and high rate production in an environmentally benign manner.

Nanotechnology is not only a frontier-field of science, but also the basis for a new economy. Nanomanufacturing is expected to provide new capabilities and product or service lines to traditional industries, such as automotive, aerospace, electronics, power, chemical, biomedical, and health. Furthermore, nanomanufacturing holds the exceptional promise of creating completely new markets for the nascent manufacturing industries of nanoparticles, nanostructures, and nanodevices.

Advances in nanomanufacturing are anticipated to result in rapid progress in nanomaterials technology, providing enabling technologies for the advancement of information and communications technology, biotechnology and medicine, photonics/electronics technology, and security. Examples include developments in high-density data storage and manipulation utilizing

Table 1. Examples of nanomanufacturing processes

Bottom-Up Processes	Top-Down Processes
Contact printing, Imprinting	E-beam, ion beam lithography
Template growth	Scanning probe lithography
Spinodal wetting/dewetting	Optical near field lithography
Laser trapping/tweezer	Femto-, atto-sec laser
Assembly and joining (Self- and directed assembly)	Material removal processes (mechanical, chemical, and hybrids)
Electrostatic (coatings and fibers)	Electro-erosive processes (electro, chemical, and mechanical)
Colloidal aggregation	Ultrasonic material removal

"nanosemiconductors," molecular electronics, and spintronics. Nanoscale structures can be used in new biomaterials for prosthetic devices and medical aids. Significant improvements in diagnostics and drug delivery can be achieved through nanostructures that can more effectively target defective cells and travel within the human body. Similarly, because of high surface area/volume ratios, safety and security aspects such as sensors, filters, and decontaminants demonstrate increased sensitivity and yield. Finally, nanoelectromechanical (NEMS) devices and nanorobots can potentially serve as both nanomanufacturing facilitators and end-products.

Of equal importance is the recognition that nanotechnology represents a whole new way of looking at technical education, cutting across traditional departmental boundaries. Current research efforts in nanomanufacturing rely on the combined expertise of scientists and engineers, including biologists, chemists, physicists, materials scientists, mechanical engineers, chemical engineers, and electrical engineers. This is expected to provide new dimensions and open new opportunities in the education business, for nurturing a new generation of nanoscience and nanoengineering researchers and teachers, and for training the nanotechnology workforce professionals. As such, new educational tools and programs will be needed to educate "nanotechnologists", i.e., researchers and operators characterized by high multi-disciplinary skills, by capitalizing on and transforming the existing academic institutions and laboratory infrastructure. Finally, as with any new technology area, it is critical that the environmental, safety, ethical, and social impacts be studied and addressed. In any engineering design, the opportunities for doing it right exist at the beginning of the process. In this new technology, the environmental considerations must be taken into account in the initial design stages.

Developments And Insight

In the following description the research areas are organized into four thematic areas: (a) nanomaterials and nanomanufacturing; (b) prototyping, scale-up, and integration issues in nanomanufacturing; (c) measurement and metrology; and (d) theory, modeling, and simulation.

Nanomaterials and Nanomanufacturing

Nanomaterials is perhaps the area where products are likely to be most quickly realized. A few commercial examples already exist, including coatings for IR and UV radiation protection or abrasion resistance (Nanophase Technologies Corp.) and "aerogel" insulation material fabricated using the sol-gel synthesis (Nanomat, Inc.). Despite the rapid growth, a thorough scientific understanding of how to optimize these materials is still lacking. Key short-term objectives include understanding the deformation mechanisms governing the interface between matrix and nanophase in *nanocomposites* and understanding fluid/surface interactions and fluid properties in *nanofluidic* applications. Similarly, while carbon nanotubes and other nanostructures have been fabricated, there is a critical need for cost-effective, high volume synthesis and processing of *nanoparticles, nanofibers, ultrathin/monolayer films* and other building block structures from a wide range of materials. To assist with printing and imprinting type processes, new *lubricant, surfactant, and resist* type materials are needed. At a higher level of complexity, development times for new processes and products could be shortened by the creation of *standard building blocks and substrates such as nanoprinted "breadboards"* to be used by researchers as a foundational component of their research or nanosystem.

Longer term, more complex biological systems serve as biomimetic models. For example, just as skin and internal cells respond continuously to a variety of external stimuli, *adaptive surfaces* can be constructed and programmed. To achieve high production and low defect rates, many of these systems may require investigation into *self-assembly by competing interactions* (e.g., block co-polymers – kinetics and thermodynamics). Self-assembly has limitations for multi-scale structures; thus, processes may evolve to incorporate molecular machines or nanorobots formed from *bi- and multi-stable molecular systems.* Nanotechnology should also play a major role in next generation catalysts. Improved catalysis is a key to better energy efficiency, reduced environmental impact, and better process economics.

Prototyping, Scale-up, and Integration

Critical to the continued investment in and support of the general public for nanotechnology is the transfer of laboratory-scale successes to the creation of commercial products. In the short term, it is imperative to be able to *produce prototype devices and examples of commercially viable manufactured products* that demonstrate nanoscale functionality. This means both large-scale processes and hierarchical assembly. At the first level, approaches are needed for the *high rate, high volume fabrication/synthesis of building blocks* (e.g., dots, wires, tubes, particles, fibers, films) from a greater range of materials and with better control of size, shape, and their polydispersity. Integrating these building blocks requires an understanding of issues such as substrate and building block surface modification for directed self-assembly. In addition to self-assembly, it is recognized that product realization may come about as a combination of new and traditional manufacturing processes, bridging top-down and bottom-up approaches. Therefore, the *integration of nanoparticle and nanomaterial synthesis with subsequent manufacturing steps* and the consolidation and forming of nanostructures into macroscopic objects will be needed. Some promising manufacturing processes that need further development include *patterning and deposition technologies* (e.g., stamping, printing), *rapid prototyping processes* (e.g., EBL), and scale-down of *lab/fab-on-a-chip* type devices. The objectives are to create sustainable, user-friendly, environmentally compatible, safe and health-preservative, affordable, high throughput, large area fabrication of a wider range of materials.

As the technology moves forward, better control over *three-dimensional assembly, interconnection* of nanostructured devices (e.g., with microcircuits), and *manipulation and rapid setup* of material components in multi-step fabrication are important goals. Reliability of nanostructures relies on control of surface/interface composition/structure to *minimize defects* and enable subsequent processing (e.g., nanoscale planarization, polishing), and on the ability to remove and repair defects in nanofabricated structures

Measurements and Metrology

To study features and phenomena at the nanoscale requires instruments capable of resolutions at the nano-, subnano-, and even pico-levels. The challenge in nanomanufacturing will be not only to develop new experimental and analytical tools with a broader range of capabilities at the nanoscale (e.g., chemical analysis, surface and sub-surface defects, sub-surface properties, charge transport, spectroscopy), but these tools must also *work in situ, real-time, non-intrusively or destructively, and under the variable conditions* seen in processing (e.g., temperature, pressure, electrical and magnetic fields). As new instruments are developed, new methods of calibration and standardization and in concert, affordable calibration standards, must be developed to ensure the accurate interpretation of results. For calibration, measurement, and assembly, reproducible positioning and repositioning with nanometer accuracy is needed.

Theory, Modeling, and Simulations

As nanotechnology builds upon the unique properties that matter exhibits in the form of small particles or structural clusters, nanomanufacturing research resides in the space between individual atomic/molecular (e.g., quantum) theories and bulk continuum theories. Because of this, many of the existing models and assumptions are no longer valid at this scale. Experimentation will be necessary to identify the reasons for which the existing theories break down, but to move forward requires the development of new theories, models, and simulations. Fundamentally, there are the basic structure-property-processing-performance relationships that must be established for nanomaterials and nanostructures. Of particular interest to nanomanufacturing is process modeling in restricted spatial domains where boundary effects become pervasive,

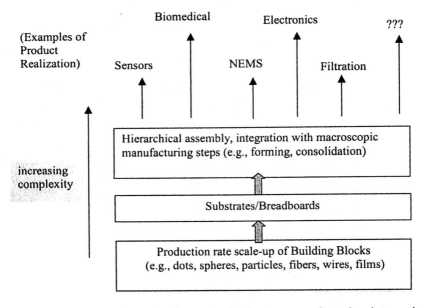

Figure 1. Schematic of levels of complexity for nanomanfacturing integration and assembly.

nanocontact mechanics, and coupled and multiscale models. These models include coupling of various spatial and time scales, as well as coupling of physical/chemical phenomena, including non-equilibrium phenomena (e.g., transport, growth). Other primary needs include developing potentials that are appropriate for nanostructure interactions, utilizing *ab initio* approaches to benchmark their accuracy, and exploring statistical mechanics approaches to thermodynamic properties and phase diagrams. Ultimately, as knowledge in this area evolves, new design paradigms will need to be created for nanomanufacturing. For example, the concept of a mechanical nanogear with mechanically meshed teeth may be replaced with interlocking motion that is governed by surface chemistry.

Security and Environmental Issues

Since September 11, 2001, there has been a heightened sense of importance and urgency in addressing international security issues. This emphasis has been on protection of the public from nuclear, chemical and biological threats. Increased public protection also leads to increased protection from pollution. Nanomanufacturing plays an important role in the mass production of affordable, fast-response, high sensitivity, portable, low power nanosensors (e.g., molecular recognition). Ultimately, these sensors can be incorporated into broadly applicable sensor arrays and actuators for both environmental and smart/defensive nanosystems and real-time, integrated sensing and chemical/biological response. Passive systems such as selectively permeable membranes will also serve to filter undesired components for applications such as security, water filtration, and dialysis.

On the structural side, lightweight/high strength/high toughness "intelligent" nanocomposites and nanocoatings can be developed for mechanical, thermal, electrical, and radiation protection. In addition, these materials can benefit the environment, for example, through dematerialization and fuel savings through decreased vehicle weight.

A second aspect involves longer-term security, which is based on health, environmental, economic, and social stability consequences of production and use of nanomanufactured products and services. As new products and processes are developed, a strong effort must be made to ensure that they are healthful, and environmentally safe and benign. Improvements in nanomanufacturing will tend to focus on making useful products mass producible and affordable to the general public. Because of the potential to build up from nanoscale, there is a great opportunity for drastic reduction in process waste generation, as well as remediation and recycle of waste materials, such as carbon particles and organic compounds. Advances in energy efficiency from nanotechnology will bring down the cost of doing business, along with reducing our reliance on non-renewable and environmentally hazardous resources.

Environmental Impact

Nanotechnology has the potential to provide environmental benefits to the community through, for example, providing means for ultra-sensitive detection of chem-bio agents and decontamination of waste, and prevention of polluting wastes in the first place. However, ultra-care is essential to explore the environmental impact of nano building blocks from the stages of manufacturing-to-integration-to-application. It is imperative to learn from the examples of technological revolutions of the last century to avoid some of the same most taken paths, which may lead to environmentally hazardous waste, and may eventually lead to tampering of biosphere. Hence, special attention is essential to generate awareness, knowledge and protocols/standards for effective and environmentally benign progress of nanomanufacturing.

Interactions With The Community At-Large

Critical to the ultimate success of the nanomanufacturing effort is being able to inform the general public about the promise of nanomanufacturing with respect to new products and services, and opportunities in education, research, and professional work. Dissemination of information to the general public must be done in a form that is exciting and understandable, utilizing widely distributed media such as popular magazines, newspapers, and TV. On a somewhat more narrow scope, informing the broader technical community of nanomanufacturing challenges, particularly with consideration of the environment, and successes should be done via special symposia at major technical meetings, joint meetings, and the establishment of a pool of distinguished speakers to present to communities not currently active in nanotechnology. These efforts will help to cross-fertilize interactions across various disciplines, and stimulate nanomanufacturing research and education involvement of scholarly and professional societies from traditional disciplinary fields.

Acknowledgements

Respectfully, we acknowledge Fabio Biscarini (ISMN, Bologna, Italy), and Carlo Taliani (ISMN, Bologna, Italy) for their partnership and discussions.

Chapter 43

Synthesis of Surfactant Templated Aerogels with Tunable Nanoporosity

Ahmad Al-Ghamdi, Sermin G. Sunol, and Aydin K. Sunol*

Chemical Engineering Department, University of South Florida, Tampa, FL 33620
*Corresponding author: sunol@eng.usf.edu

Several varieties of regular and surfactant-templated alumina aerogels and xerogels were synthesized. The problems associated with retention of textural properties were greatly reduced through surfactant-templated sol-gel synthesis pathways for alumina and subsequent removal of the solvents and templates from the gel network by a technique that involves a supercritical drying method. For the first time, Surfactant-Templated Alumina Aerogel (STAA) was made and showed remarkable thermal stability and gave specific surface area of as high as 1000 m^2/g for as-synthesized STAA and up to 600 m^2/g for calcined STAA. Extent of removal of surfactant was determined using FTIR. Furthermore, NiO/Al_2O_3 catalysts were prepared by co-precipitating the active metal and support. Nitrogen adsorption desorption isotherms (NAD) were used to follow the evolution of the extraction of the pore solvents and templates and to follow the textural properties and XRD was used to follow the morphological properties and the structural phase change for the synthesized gels. SEM-EDS is utilized to study dispersion of nickel on alumina.

Introduction

Aerogels were first synthesized by Kistler in 1931 [1]. They are highly porous material formed when precursor gels are dried at temperatures and pressures higher than the critical point of the pore fluid. The hypercritical drying process replaces the original pore liquids with supercritical fluid. Thus, high capillary stresses, associated with conventional drying, that result in collapsed pores are avoided. The hypercritical drying processes result in little shrinkage associated with drying. Aerogels may be used as catalysts, thermal insulators, chemical adsorbents, sensors, fuel storage devices, energy absorbers and aero capacitors [1-5]. Some of their properties include high porosity, large pore volume, high surface area and morphological stability at high temperatures [6]. The sol-gel method offers several advantages for making aerogels, such as high purity, microstructure, homogeneity at molecular level, and low temperature at preparation [5, 7, 8]. After the discovery of M41S family of solids by Beck et al. in 1992 [9], the new pathway have been extended to other porous base materials, such as alumina, and to have better control over the properties of the porous materials. This achieved through use of surfactants in the synthesis of the sol-gel [10]. The advantages of the surfactant-templated porous materials include high porosity and tunable unimodal nanoporosity. However, the main disadvantage of surfactant-templated xerogel is that they undergo 75% to 95% volume shrinkage during drying and calcination [11, 12]. In this paper, we present a methodology and results for thermally stable surfactant-templated alumina aerogels and NiO/Al_2O_3 catalysts. The scope of the paper includes the synthesis of templated and non-templated aerogels, their calcination and textural and morphological characterization of the material and metal dispersion for NiO/Al_2O_3 catalysts. For comparison purposes, xerogels made of the same materials were prepared and characterized.

Experimental

Preparation of Nickel-Alumina gel

The desired amount of the surfactant Triton X-114 (Sigma) was dissolved in the desired amount of the *sec*-butanol (Aldrich). Then, Aluminum secbutoxide ASB (Aldrich) was added to the solution. Nickel acetate was dissolved in methanol and was mixed with the required amount of distilled water. The first solution was added to the second one with continuous stirring for one hour at room temperature. The mixture was then further stirred for one hour at room temperature. Then the gel was stored in its own solvents at room temperature for two days for aging. The aged sol-gel sample was then washed with ethanol (Aldrich) before it was extracted in a Soxhlet extractor with ethanol for 12 hours. The molar ratio for making the sol-gel was ASB:*sec*-butanol:X:H2O are 1:20:0.1:2 respectively, where, X is surfactant triton X-114. However, other surfactants concentrations were also used.

Drying of Alcogels

For making xerogel, the alcogel was allowed to dry at ambient pressure and 313K for three days. For making aerogel, the alcogel was loaded into the autoclave for CO_2 supercritical drying. The autoclave is heat-jacketed and the temperature was set to 333K. CO_2 at 1800 psi pressure, was pumped to the autoclave in a continuous mode during the entire supercritical drying process. A recirculation system that permits high recycle ratios and superficial gas velocities make the supercritical drying process faster and solvent efficient.

Calcination

For calcining the xerogels and aerogels, samples were heat treated at different temperatures 773K, 873K, 973K in a flowing air atmospheric pressure.

Characterization of Samples

Surface area, porosity and adsorption-desorption isotherms were obtained using a Nova 3000 apparatus. XRD is used to analyze alumina and aluminum hydroxide phases. Bulk densities are determined gravimetrically. SEM-EDS analysis is used to determine Nickel distribution on alumina.

Results

Pore Volume and Pore Size Distribution

The pore volumes of Alumina gels as determined from the desorption part of the NAD isotherms using BJH method and are shown in Table 1. Unlike all other samples, ST xerogel showed a consistent trend of gaining pore volume when it was calcined at the first temperature (773K), which suggests that ST xerogel still has considerable amount of the templates within the pore mesostructure before calcinations, even though it was Souxlet extracted and dried for several days. On the other hand, in the case of the ST aerogel, there was no gain in the pore volume after calcination, which suggests that supercritical CO_2 drying removes additional amount of the templates left after the Souxlet extraction. The average pore size measured by NAD is also shown in Table 1. ST aerogel has the largest average pore size. ST xerogel and regular aerogel have average pore size close to the ST aerogel. Regular xerogel has a pore size distribution closer to micropores.

BET Specific Surface Area

The effect of calcination temperature on specific surface area of the alumina aerogel and xerogel is shown in Table 1 for both surfactant-templated (ST) and regular (prepared without use of any surfactant) aerogels and xerogels. ST aerogel shows high surface area and high thermal stability. ST xerogel has considerable less specific surface area than ST aerogel, yet, the area is still higher than surface areas of both regular aerogel and regular xerogel. Table 1 shows the textural properties for the prepared samples including specific surface area values of about 1000 m^2/g and 700 m^2/g for as synthesized ST aerogel and ST xerogel respectively.

Bulk Densities

Figure 1 shows the bulk density as a function of the surfactant (X-114) concentration used in the sol gel initial concentration. It is apparent that ST aerogel poses the lowest bulk density. The calcined ST aerogel showed more stable density after calcination at X-114 concentration of 0.1 and seems to undergo more volume shrinkage for the other surfactant concentrations. The

Table 1 Textural properties for templated and regular gels

Temperature, K	ST Aerogel			ST Xerogel			Regular Aerogel			Regular Xerogel		
	Sa, m²/g	Vp, cc/g	Dp, A	Sa m²/g	Vp cc/g	Dp A	Sa m²/g	Vp cc/g	Dp A	Sa m²/g	Vp cc/g	Dp A
As-Synthesized	997	2.56	103	587	0.61	41	763	0.77	40	535	0.82	61
773	701	2.15	123	533	0.94	71	503	0.62	50	439	0.81	74
873	631	1.94	123	483	0.79	65	455	0.53	47	404	0.75	75
973	610	1.63	107	442	0.77	69	373	0.44	48	371	0.66	71

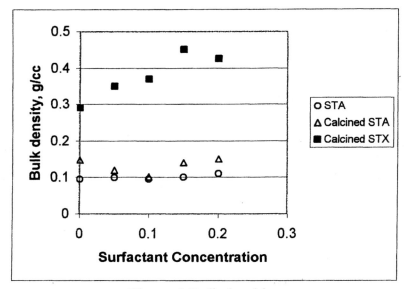

Figure 1 Bulk densities.

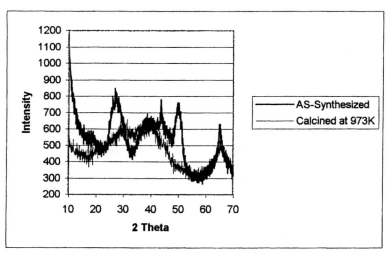

Figure 2 X-ray diffraction patterns for ST aerogel, as-synthesized and calcined at 973K.

sample processing steps (drying and calcination) account for volume shrinkage of 5 and 75% and porosity was 98 and 81% for the ST aerogel and ST xerogel respectively.

XRD Phase Characterization

Figure 2 shows the phase progress for the ST aerogel. The as-synthesized ST aerogel shows a diffraction pattern that was assignable to boehmite alumina structure. Sample calcined at 973K mainly had an amorohous diffraction pattern plus a little diffractions that could be assigned to gamma phase alumina. NiO-Alumina samples show peaks that can be assigned to $NiAl_2O_4$(Figure 3).

Nickel Dispersion

SEM-EDS analysis of co-precipitated NiO/Al_2O_3 samples show good dispersion of nickel on the alumina support. Ni/Al mass ratio on the surface is 4.1%, as determined by SEM/EDS. The mass of Nickel introduced 4.9%.

Conclusion

Thermally stable alumina supports are synthesized using surfactant templating and supercritical drying. Co-precipitation permits even and high dispersion of the metal on the support when NiO/Al_2O_3 catalysts are prepared using surfactant templating and supercritical drying.

References

[1] Kistler, S.S., Nature, Vol. 127 1931, P.741
[2] Hrubesh, L.W., Journal Of Non-Crystalline Solids, Vol. 225, 1998, P.335
[3] Knozinger, H., Ratnasamy, P., Catal. Rev. Sci. Eng, Vol. 17 (1), 1978, P.31
[4] Brinker, C.J., Scherer, G.W. "Sol-Gel Science The Physics And Chemistry Of Sol-Gel Processing", 1990, Academic Press.
[5] Suh, D.J., Journal Of Non-Crystalline Solids, Vol. 225 1998, P.168

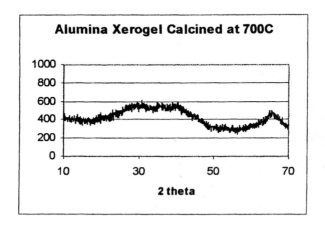

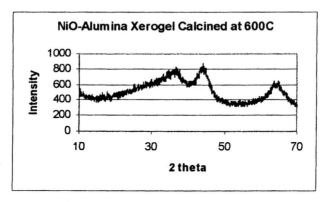

Figure 3. XRD of Alumina and NiO-Alumina Xerogels

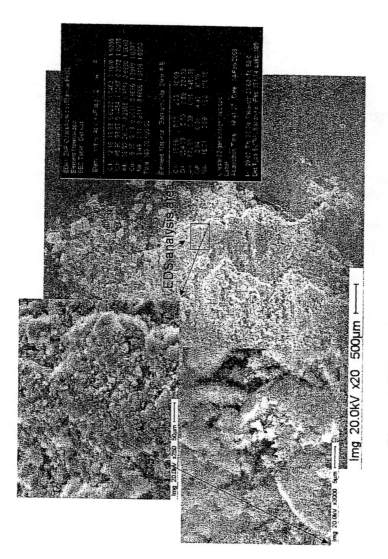

Figure 4. SEM images of NiO/Al$_2$O$_3$ catalysts

[6] Janosovits, U., Ziegler, G., Scharf, U. Wokaun, A., Journal Of Non-Crystalline Solids, Vol. 210, 1997, P.1

[7] Ertl, G., Knozinger, H., Weitkamp, J., "Preparation Of Solid Catalysts", 1999 Wiley-Vch

[8] Bagwell, R.B., Messing, G.L., Key Engineering Materials, Vol.115 1996, P.45

[9] Beck, J.S., Vartuli, C., Roth, W.J., Leonowicz, M.E., Kresge, C.T., Achmitt, K.D., Chu, C.T-W., Olson, D.H., Sheppard, E.W., Mccullen, S.B., Higgins, J.B., Schlenker, J.L., J. Am. Chem. Soc. Vol.114, 1992, P.10834.

[10] Gonzalez-Pena, V., Diaz, I., Marquez-Alvarez, C., Sastre, E., Perez-Pariente, J., Microporous And Mesoporous Materials Vol. 44-45, 2001, P.203

11- Mark T. Anderson, Patricia S. Sawyer, Thomas Rieker, Microporous And Mesorporous Materials Vol. 20, 1998, P.53-65.

12- Mark .T. Anderson, J.E. Martin, J.G. Odinck, P.P. Newcomer, J.P. Wilcoxon, Microporous Mater. Vol.10, 1997, P.13

Environmental Applications: Nanotechnology-Enabled Green Energy and Power

Editor
Debra R. Rolison

Chapter 44

Energy and the Environment: Perpetual Dilemma or Nanotechnology-Enabled Opportunity?

Debra R. Rolison

Surface Chemistry Branch, Code 6170, Naval Research Laboratory, 4555 Overlook Avenue, SW, Washington, DC 20375 (rolison@nrl.navy.mil; fax: 1–202–767–3321)

The global demand for the energy that sustains human-based activity is unceasing—and increasing. This global need drives production processes (including extraction, chemical manipulation, and distribution) and in-use processes that potentially can compromise environmental quality. Yet the global and local environment can only be sustained, cleaned, and preserved through the expenditure of energy. A perpetual irony it may be, but thermodynamics demands that truism.

Are there environmentally green opportunities foreseeable by re-thinking and re-designing energy production and power generation from a nanoscopic perspective? The heterogeneous catalysis necessary to process petroleum into liquid transportation fuels is (and has always been) innately nanoscopic *(1)*. Even with a history that long precedes today's focus on nanoscale science and engineering, improved catalytic chemistries are still sought, particularly to desulfurize fuels, in order to lower the environmental impact of the use of fossil fuels *(2)*.

Breakthroughs in energy storage and conversion are arising from the confluence of nanoscale S&T, environmentally green considerations, and the importance of alternate and distributed energy. Multifunctional materials are prerequisite to electrochemical and thermal-to-electric power sources. In order to deliver high performance in such devices, multifunctional materials must exhibit some combination of the following properties: electronic conductivity, ionic conductivity, thermal conductivity, separation of electron–hole pairs, catalytic selectivity and activity, as well as facile mass transport of molecules in the electrochemical devices.

Independent control of the elementary processes that give rise to energy-relevant functionalities is difficult-to-impossible with bulk materials. Scientists and engineers who design, fabricate, and study matter on the nanoscale are recognizing that an obvious research opportunity lies in determining how to assemble nanoscopic building blocks into multifunctional architectures that store or convert energy and produce power (3-6).

Another critical design parameter arises from the need to move beyond the high level of understanding that has been developed for electron-transfer reactions between molecules or between electrodes and molecules (7) and to understand charge transfer on the nanoscale (8). This knowledge base is also critical to advances in molecular electronics (9). Progress in the areas of nanomaterials and nanoarchitectures, and new conceptual insights into charge transfer in molecular wires will ultimately contribute to the design of improved devices for energy storage and conversion.

The electrified interfaces that power electrochemical storage and conversion devices, such as batteries, fuel cells (including biofuel cells), electrochemical capacitors, and dye-sensitized photovoltaic cells, require the motion of mobile ions in the electrolyte to counterbalance surface charges that arise as the energy of the electrons (i.e., the potential) varies. When the electrode becomes nanoscopic, and especially when it contacts a volume of electrolyte or fuel/oxidant in a nanoscale pore, size matters. One well-studied porous electrode, carbon, alerts those interested in devising nanostructured electrochemical power sources that the critical factor is not high surface area, but rather molecule-, ion-, and solvent-accessible surface area on the time scale of the electrochemical processes that store or generate ionic and electronic charge (10,11). In particular, the surface area that resides behind pore openings sized at ≤ 1.5 nm, such as is typical in single-walled carbon nanotubes does not contribute to the electrochemically active surface area.

Mesoporous nanostructured materials comprising bicontinuous networks of porosity and electrochemically active nanoscopic solid have already demonstrated the importance of arranging nanoscale building blocks in space to create multifunctional power-source architectures (3). Sol–gel-derived materials, such as aerogels, in which the porous network retains through-connectivity (by minimizing compressive forces during processing), provide a means to investigate how electrochemical processes are influenced by (1) minimal solid-state diffusion lengths; (2) effective mass transport of ions and solvent to the nanoscale, networked (indeed, self-wired) electrode material; and (3) amplification of the surface (and surface-defect) character of the electrochemically active material.

Recent work with nanoscopic, highly mesoporous, networked charge-insertion oxides has indicated that such architectures store electrical charge by three, not two, means: (1) double-layer capacitance (as measured using bulky, non-insertion cations in which normalizing the capacitance by the physisorption-derived surface area yields standard values); (2) pseudocapacitance readily accessible at high rates of discharge (ideally expressed as a constant $\partial C/\partial E$, with charge stored over a range of potentials atypical of a localized faradaic redox process); and (3) energetic-specific insertion of cations, as would be seen in a standard insertion battery material. These "colors" of capacitance at cation-insertion materials are depicted schematically in Figure 1.

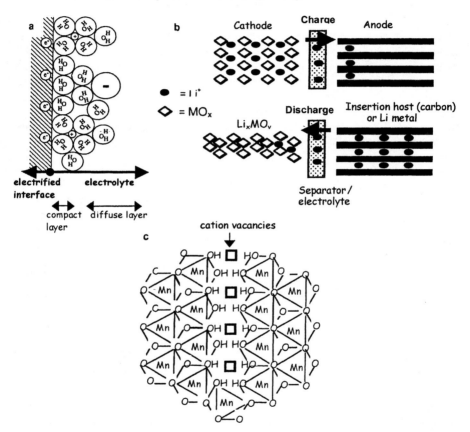

Figure 1. *The three types of charge stored at an electrified cation-insertion material; (a) formation of double-layer capacitance at an electrified interface (in this scheme at a potential negative of the point of zero charge); (b) electrochemical intercalation of Li^+ into a battery cathode; (c) proposed cation vacancies and accommodation of vacancy-balancing protons in MnO_2. (Reproduced from reference 5. Copyright 2001 Royal Society of Chemistry.)*

Materials that blend all three mechanisms of charge/energy storage are innately hybrid and should be able to provide future functions now served by the physical/electrical coupling of battery and capacitor in applications that require both peak power and high energy density expressed as low levels of power sustained over long duration use *(5)*. The ability to store both battery-like charge and ultracapacitor-like charge is attributed to the disorder present in the nanoscale materials, which derive from sol–gel processing *(5)*. Inducing proton-stabilized cation vacancies in polycrystalline, 1-μm-sized V_2O_5 powder (by heating at 460 °C under an oxygen/water atmosphere) was recently shown to increase the Li-ion capacity of the oxide relative to the as-received material, while inducing anion vacancies (by heating at 460 °C under a low partial pressure of oxygen) lowered the lithium-ion capacity *(12)*. Coupling an ability to make disordered nanoscopic insertion materials with a computational understanding of how to increase the number of electrons transferred per transition metal cation, while minimizing deleterious structural or phase changes in the insertion material *(13)*, will spur improvements in batteries impossible to achieve with mesoscale materials and structures.

The energy conversion realized by fuel cells is innately nanoscopic, because the electroreactions in H_2/O_2 or direct methanol fuel cells are catalyzed, usually by carbon-supported nanoparticulate Pt-group-metals. But rather than just resolve the demands of standard heterogeneous catalysis (activity, selectivity, facile transport of reactants to and products from the catalytic site), the electrocatalysis that is essential in a fuel cell adds the need for high mobility of electrons and ions to and from the catalytic site. These additional demands create what is known as the three-phase boundary *(14)* between *(i)* the solid electrocatalyst (perched on or nestled in an electron-conductive porous carbon, which serves as the current collector that transports electrons—the true reactants and products of the fuel cell), *(ii)* the molecular fuel or oxidant (frequently a gas), and *(iii)* the electrolytic medium that is liquid, gel-like, or polymeric and contains solvated mobile ions. The only truly reactive zone is the point (or line) of contact at the junction of the three phases (see Figure 2 *(15)*). Even with decades of work on this issue, fuel cells still require the design of improved structures to maximize the effective area of the three-phase boundary as well as the transport of all species to and from it. These challenges are opportunities awaiting creative nanoarchitectural design *(4)*.

Advances in nanoscale science and technology will improve the nature of the electrocatalysts and architectures critical to high-performance fuel cells, but nanoS&T already plays a key role in electrochemical devices that convert solar energy into electrons (e.g., in the dye-sensitized, wide-bandgap nanocrystalline semiconductor photovoltaic cells *(3)*), or solar energy into fuel *(16)*, or in the control of enzymatic or biomimetic catalysis in order to convert non-petroleum-derived fuels into electron energy *(17)*. In the realm of thermal/electric, rather than chemical/electric, controlled fabrication on the nanoscale of thin-film superlattices of known thermoelectric materials has broken decades of stalemate to lead to improved thermoelectric performance relative to the performance of devices using the bulk material *(18,19)*.

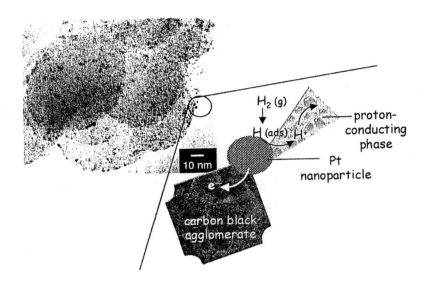

Figure 2. *The three-phase boundary phase at a carbon-supported Pt electrocatalyst. To electrogenerate power (electrons per second at a specific potential) the reactive interphase requires transport of a gas-phase reactant (H₂ in this example) to and adsorption at a Pt nanoparticle while maintaining intimate contact to a proton-conducting phase (such as an acid electrolyte or a proton-exchange polymer). (Micrograph is reproduced from reference 15. Copyright 2002 American Chemical Society.)*

Future contributions to the materials of importance in energy storage and conversion will also require better theoretical, synthetic, and characterization approaches to disordered materials. A key opportunity/challenge will be to "pin" the most active, highest performance physicochemical state of the material even when exposed to thermodynamic forces (temperature, pressure, electrical potential, photochemical energy) that would otherwise drive restructuring, crystallization, denaturation, or densification of the nanoscale energy-storage or –conversion material.

The session on "Nanotech-enabled green energy sources," held as part of the symposium on "Nanotechnology and the Environment" at the 225th Meeting of the American Chemical Society, 23–27 March 2003 in New Orleans, explored electrical energy as derived from electrochemical, thermal, solar, and biological conversion processes from the perspective of re-design and optimization on the nanoscale. Six experts described their latest work on the importance of nanoscience on fundamental processes that produce energy. The papers derived from this session follow and explore *(i)* the nature of charge transfer on the nanoscale; *(ii)* mesoporous battery nanoarchitectures; *(iii)* nanostructured thin-film superlattices in thermoelectric devices; *(iv)* membraneless, microfluidic-based biofuel cells; and *(v)* hybrid energy converters that couple dye-sensitized, wide-bandgap semiconductor photovoltaics and biofuel cells.

The fundamental science in all these areas is well underway. The future research will be richly influenced by nanoS&T, inevitably yielding new design strategies for nanotech-enabled green energy.

Acknowledgements

The author is grateful for the sustained support of her team's research on multifunctional nanoarchitectures by the Office of Naval Research and the Defense Applied Research Projects Agency (DARPA).

References

1. Kaufmann, T. G.; Kaldor, A.; Stuntz, G. F.; Kerby, M. C.; Ansell, L. L. Catalysis science and technology for cleaner transportation fuels. *Catal. Today* **2000**, *62*, 77–90.
2. Rossini, S. The impact of catalytic materials on fuel reformulation. *Catal. Today* **2003**, *77*, 467–484.
3. Grätzel, M. Photoelectrochemical cells. *Nature* **2001**, *414*, 338–344.
4. Rolison, D. R. Catalytic nanoarchitectures—the importance of nothing and the unimportance of periodicity. *Science* **2003**, *299*, 1698–1701.
5. Rolison, D. R.; Dunn, B. Electrically conductive oxide aerogels: new materials in electrochemistry. *J. Mater. Chem.* **2001**, *11*, 963–980.
6. Cava, R. J.; DiSalvo, F. J.; Brus, L. E.; Dunbar, K. R.; Gorman, C. B.; Haile, S. M.; Interrante, L. V.; Musfeldt, J. L.; Navrotsky, A.; Nuzzo, R. G.; Pickett, W. E.; Wilkinson, A. P.; Ahn, C.; Allen, J. W.; Burns, P. C.; Ceder, G.; Chidsey, C. E. D.; Clegg, W.; Coronado, E.; Dai, H. J.; Deem, M. W.; Dunn, B. S.; Galli, G.; Jacobson, A. J.; Kanatzidis, M.; Lin, W. B.; Manthiram, A.; Mrksich, M.; Norris, D. J.; Nozik, A. J.; Peng, X. G.; Rawn, C.; Rolison, D.; Singh, D. J.; Toby, B. H.; Tolbert, S.; Wiesner, U. B.; Woodward, P. M.; Yang, P. D. Future directions in solid state chemistry:

330

report of the NSF-sponsored workshop. *Progress Solid State Chem.* **2002**, *30*, 1–101.

7. Jortner, J.; Bixon, M. Electron transfer–from isolated molecules to biomolecules. *Adv. Chem. Phys.* **1999**, *106*, 35–202, and references therein.
8. Adams, D. M.; Brus, L.; Chidsey, C. E. D.; Creager, S.; Creutz, C.; Kagan, C. R.; Kamat, P.; Lieberman, M.; Lindsay, S.; Marcus, R. A.; Metzger, R. M.; Michel-Beyerle, M. E.; Miller, J. R.; Newton, M. D.; Rolison, D. R.; Sankey, O.; Schanze, K. S.; Yardley, J.; Zhu, X. Charge transfer on the nanoscale: current status. *J. Phys. Chem. B* **2003**, *107*, 6668–6697.
9. Nitzan, A.; Ratner, M. A. Electron transport in molecular wire junctions. *Science*, **2003**, *300*, 1384–1389.
10. Koresh, J.; Soffer, A. Double-layer capacitance and charging rate of ultramicroporous carbon electrodes. *J. Electrochem. Soc.* **1977**, *124*, 1379–1385.
11. Yang, K. L.; Yiacoumi, S.; Tsouris, C. Electrosorption capacitance of nanostructured carbon aerogel obtained by cyclic voltammetry. *J. Electroanal. Chem.* **2003**, *540*, 159–167.
12. Swider-Lyons, K. E.; Love, C. T.; Rolison, D. R. Improved lithium capacity of defective V_2O_5 materials. *Solid State Ionics* **2002**, *152–153*, 99–104.
13. Hwang, B. J.; Tsai, Y. W.; Carlier, D.; Ceder, G. A combined computational/experimental study on $LiNi_{1/3}Co_{1/3}Mn_{1/3}O_2$. *Chem. Mater.* **2003**, *15*, 3676–3682.
14. Bockris, J. O'M.; Reddy, A. K. N. *Modern Electrochemistry 2*; Plenum Press: New York, 1970; pp 1382–1385.
15. Anderson, M. L.; Stroud, R. M.; Rolison, D. R. Enhancing the activity of fuel-cell reactions by designing three–dimensional nanostructured architectures: catalyst-modified carbon-silica composite aerogels. *Nano Lett.* **2002**, *2*, 235–240 [correction: *Nano Lett.* **2003**, *3*, 1321].
16. Millsaps, J. F.; Bruce, B. D.; Lee, J. W.; Greenbaum, E. Nanoscale photosynthesis: Photocatalytic production of hydrogen by platinized photosystem I reaction centers. *Photochem. Photobiol.* **2001**, 630–635.
17. Palmore, G. T. R.; Whitesides, G. M. Microbial and enzymatic biofuel cells. In *Enzymatic Conversion of Biomass for Fuels Production*; Himmel, M. E., Baker, J. O., Overend, P., Eds.; Am. Chem. Soc. Symp. Ser. 566: Washington, DC, 1994; pp 271–290.
18. Venkatasubramanian, R.; Siivola, E.; Colpitts, T.; O'Quinn, B. Thin-film thermoelectric devices with high room-temperature figures of merit. *Nature* **2001**, *413*, 597–602.
19. Harman, T.; Taylor, P. J.; Walsh, M. P.; LaForge, B. E. Quantum dot superlattice thermoelectric materials and devices. *Science* **2002**, *297*, 2229–2232.

Chapter 45

Electrochemistry at Nanometer Length Scales

John J. Watkins, Chett J. Boxley, and Henry S. White*

Department of Chemistry, University of Utah, Salt Lake City, UT 84112
*Corresponding author: email: white@chem.utah.edu;
fax: 1–801–585–5720

We briefly describe two sets of electrochemical experiments whose success and interpretation require explicit consideration of the role of distance at nanometer length scales. In the first section, measurements of the influence of adsorbed Cl⁻ on the dissolution rate of the ~2-nm-thick Al_2O_3 film on Al are reported, and analyzed in terms of the electric field across the oxide film. Knowledge of the potential-dependent film thickness, which varies by less than 1 nm during an experiment, is required to quantify dissolution rates. In the second section, we describe experiments using Pt electrodes to voltammetrically detect very small quantities (zeptomoles) of an electroactive species. Our strategy to detect such small amounts of analyte is based on shrinking the electrode size to nanometer length scales.

The Role of Nanoscale Structures in Oxide Films

The stability of many metals is due to the native surface oxide that provides a kinetic barrier against oxidative dissolution. Titanium and Al are protected by native films of TiO_2 and Al_2O_3, respectively, typically between 2 and 3 nm in thickness, that act as barriers to electron and ion transfer between the environment and the metal. The passivity provided by the oxide layers may be compromised by defect structures in the oxide film that accelerate these transport processes, and by chemical reactions that lead to oxide film dissolution. Detailed descriptions of these phenomena have eluded scientists for decades, in part, due to an incomplete understanding of the dominant role of nanometer-scale structures in the oxide film, and, in part, due to the analytical challenges in measuring chemical reaction rates and structures for such small quantities of material.

The thickness of the Al_2O_3 film on Al is determined by a dynamic balance of oxide growth, eq 1, and oxide dissolution, eq 2. The presence of Cl^- in the solution accelerates Al_2O_3 film dissolution, leading to oxide film breakdown and corrosion.

$$2\,Al + 3\,H_2O \rightarrow Al_2O_3 + 6\,H^+ + 6\,e^- \tag{1}$$

$$Al_2O_3 + 6\,H^+ \rightarrow 2\,Al^{3+} + 3\,H_2O \tag{2}$$

Figure 1 demonstrates this behavior. In the absence of Cl^-, the voltammetric response of an Al electrode displays a current that reflects irreversible oxide growth (eq 1). When Cl^- is added to the solution, the voltammetric currents increase due to Al_2O_3 dissolution (eq 2). Eventually the underlying Al is exposed to the solution, leading to rapid metal dissolution. Remarkably, in spite of the technological importance of Al, the rate of Al_2O_3 native film dissolution in the presence of Cl^- has not been established.

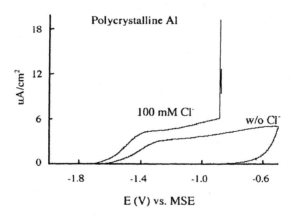

Figure 1. Voltammetric response of an Al electrode cycled at 1 mVs^{-1} in an unstirred 50 mM H₃BO₃/0.3 mM Na₂B₄O₇ • 10 H₂O solution in the absence and presence of 100 mM NaCl. The large increase in current at ca. −0.85 V occurs when local dissolution of the oxide film results in exposure of the underlying Al to the solution. (Reproduced from reference 1. Copyright 2003 Electrochemical Society.)

We have analyzed the growth and dissolution behavior of Al_2O_3 films *(1)* using a voltammetric method recently described by Isaacs and coworkers *(2)*.

A basic assumption underlying the analysis is that oxide film growth is governed by high-field transport of ions within the oxide film, as illustrated in Figure 2. In the high-field model, the electrical field determines the rate of ion transport across the oxide film. The voltammetric current due to oxide growth, i, is given by eq 3,

$$i = Nvq\exp\left[-\frac{W}{kT} + \frac{qa(E - E_o)}{kTD}\right]$$

(3)

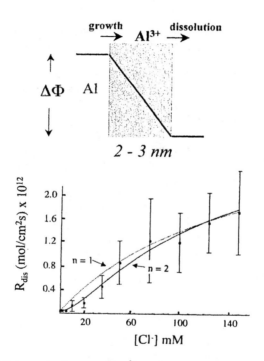

Figure 2. *(Top) Schematic diagram of Al^{3+} transport through a native oxide film on Al. $\Delta\Phi$ is the potential drop across the oxide film. The relative rates of oxide growth (an electrochemical reaction, eq 1) and oxide dissolution (a chemical reaction, eq 2) determine the thickness of the passive film. (Bottom) Dissolution rate of the Al_2O_3 film (R_{dis}) as a function of the solution concentration of Cl^-. Lines correspond to first-order ($n = 1$) and second-order ($n = 2$) dissolution kinetics. Error bars correspond to 2σ (5 independent measurements). Negligibly small increases in R_{dis} are observed upon addition of Br^- (Reproduced from reference 1. Copyright 2003 Electrochemical Society.)*

where E is the applied potential, E_o is the flatband potential, N is the surface density of mobile Al^{3+}, v is a vibrational frequency associated with ion motion in the oxide lattice, q is the charge associated with Al^{3+}, W is the activation energy, a is the activation distance, k is the Boltzmann constant, and T is temperature. In eq 3, D (cm) represents the thickness of the oxide layer, defined at any instant in time as:

$$D = D_o + \frac{M}{\rho}((nF)^{-1}\int idt - \int R_{dis}dt) \tag{4}$$

In eq 4, D_o is the initial thickness of the oxide film ($\sim$2.6 nm), M is the molecular weight of Al_2O_3, ρ is the density of amorphous Al_2O_3, n is the number of electrons transferred in producing an Al^{3+} ion, F is Faraday's constant, and R_{dis} is the rate of Al_2O_3 dissolution at the oxide film/electrolyte interface. The dissolution rate is assumed to be independent of the electric potential distribution at the interface but dependent on Cl^- concentration.

We have determined R_{dis} by fitting eq 3 and eq 4 to the voltammetric curves, as detailed in reference 1. The dependence of R_{dis} on Cl^- concentration is shown in Figure 2. R_{dis} increases from a value of 3.5 ($\pm$0.1) $\times$ 10^{-14} mol s^{-1}cm^{-2} in the absence of Cl^- to a value of 1.6 ($\pm$0.7) $\times$ 10^{-12} mol s^{-1}cm^{-2}, a nearly 50-fold increase in dissolution rate.

At Cl^- concentrations above 5 mM, R_{dis} increases steadily and appears to level off, reminiscent of an adsorption-controlled process. In our analysis, we assume that R_{dis} is proportional to the surface coverage of adsorbed Cl^-, Γ_{Cl^-}, and a chemical rate constant, k, eq 5. The exponent n in eq 5 allows for the possibility that R_{dis} might have either a zero-, first-, or second-order, etc., dependence on Γ_{Cl^-}

$$R_{dis} = (\Gamma_{Cl^-})^n k \tag{5}$$

Using a simple competitive Langmuir adsorption isotherm, we have fit the preliminary data in Figure 2 to a dissolution mechanism in which two Cl^- are involved in the rate-determining step of oxide film dissolution. More recent studies in our laboratory using single crystal Al electrodes have shown that R_{dis} is a function of the Al surface crystallography.

Voltammetric analysis provides an extremely sensitive method to measure dissolution rates of oxide films. For instance, approximately 0.5 nm of oxide dissolves during the course of the voltammetric measurement in a 100 mM Cl^- solution. In the absence of Cl^-, a negligibly small (*but measurable*) amount ($\sim$0.014 nm) of Al_2O_3 dissolves into solution. For comparison, the unit cell dimensions of Al_2O_3 are $a = 0.48$ nm and $c = 1.30$ nm, significantly larger than the measured change in thickness resulting from oxide dissolution.

In addition to dissolution reactions, passivity is also compromised by defect structures that provide electrical conductivity across the oxide film. We have shown in previous reports that microscopic sites of high electrical conductivity may act as precursor sites for breakdown of the oxide film on Ti and Al. Recently, in conjunction with C. Gardner and J. Macpherson (University of Warwick), we have employed conductivity atomic force microscopy to map the electrical conductivity of the native oxide film on Ti with high spatial resolution *(3)*. For example, Figure 3 shows deflection mode and conductivity images of a polycrystalline Ti electrode, in which high electrical conductivity in the oxide film is clearly observed along grain boundaries of the Ti substrate. The high electrical conductivity at the grain boundaries is most likely due to increased disorder in the oxide film.

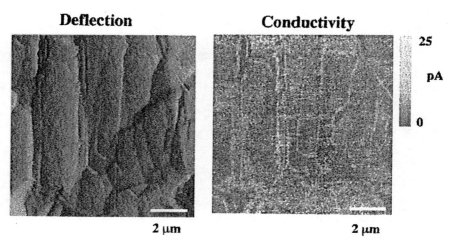

Deflection   **Conductivity**

25

pA

0

2 μm   2 μm

Figure 3. Deflection-mode (left) and conductivity (right) images of a Ti/TiO_2 surface showing the grain structure of Ti and the variation in electrical conductivity in the native oxide film at grain boundaries. (Reproduced from reference 3. Copyright 2003 American Chemical Society.)

Detection of Zeptomole Quantities of Redox-Active Molecules using Electrodes of Nanometer Dimensions

Metal electrodes with characteristic dimensions approaching 10 nm or less have been reported by several laboratories during the past 15 years *(4-6)*, and applied in a wide range of studies, including investigations of the influence of the electrical double layer on molecular transport and kinetics *(7)*, measurement of fast electron-transfer rates, and for high-resolution electrochemical imaging. It is possible to imagine using such small electrodes for quantitative voltammetric analysis of very small numbers of molecules, similar to the approach of using focused lasers to probe tiny solution volumes containing one or several fluorescent molecules. Fan and Bard have used scanning electrochemical microscopy to trap and detect single redox active molecules between a nanometer-size metal tip and a large surface *(8)*, but this approach does not readily lend itself to general analytical applications.

In recent studies *(4,5)*, we have prepared quasi-hemispherical Pt electrodes with radii between 5 and 100 nm that exhibit well-behaved voltammetric responses for the oxidation or reduction of soluble redox species. These electrodes are constructed using an electrophoretic polymer coating method originally developed by Bach et al. for coating STM tips *(9)*, and adapted by Schulte and Chow *(10)* for preparing carbon-fiber microelectrodes. Slevin et al. demonstrated the use of this method in preparing nanometer-scale Pt electrodes *(11)*. The electrodes are very stable in aqueous solutions and can be characterized by electron microscopy and various electrochemical techniques.

The voltammetric response of a Pt electrode corresponding to the oxidation of 2 mM $FcTMA^+$ in 0.2 M KCl, and a transmission electron microscopy (TEM) image of the same electrode, are shown in Figure 4. The electrochemical response is ideal, showing little hysteresis on the return scan. The apparent electrochemical radius of the electrode is calculated from the steady-state limiting current, i_{lim}, assuming a hemispherical geometry.

$$i_{lim} = 2\pi n F D C^* r_{app} \qquad (6)$$

In eq 6, C^* and D are the bulk concentration (2 mM) and diffusivity of $FcTMA^+$, respectively, r_{app} is the apparent radius, and n is the number of electrons transferred per molecule. The calculated radius is referred to as an apparent radius because an ideal hemispherical electrode geometry is assumed. A value of $r_{app} = 70$ nm was computed from the voltammetric response ($i_{lim} = 64$ pA). The geometric radius of the electrode in Figure 1, measured from the TEM image, is ~75 nm, in good agreement with the electrochemically measured radius, r_{app}, demonstrating that the electrochemical response is consistent with the observed geometry.

How few molecules can be detected at electrodes as small as the one shown in Figure 4? This question can be addressed by a variety of methods. Our

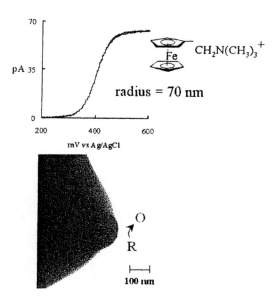

Figure 4. *(Top) Voltammetric response of a Pt electrode (r_{app} = 70 nm) in an aqueous 2 mM FcTMA$^+$ solution containing 0.2 M KCl as the supporting electrolyte. Scan rate = 10 mVs^{-1}. (Bottom) Transmission electron micrograph of the same electrode at 100,000× magnification. The protruding hemispheroid is the Pt electrode with r = 77 nm. (Reproduced from reference 4. Copyright 2003 American Chemical Society.)*

preliminary approach is to adsorb a well-characterized redox species, [Os(bpy)$_2$(dipy)Cl]$^+$, onto the surface of the electrode and to use cyclic voltammetry to measure the charge associated with the oxidation of the molecule. Abruña and coworkers, have reported a saturation coverage of [Os(bpy)$_2$(dipy)Cl]$^+$ at Pt of 1.1×10^{-10} mol cm^{-2}, independent of surface area.*(12)*. The number of molecules adsorbed at the Pt surface is obtained by measuring the electrical charge, Q, associated with the 1-electron oxidation of the adsorbed [Os(bpy)$_2$(dipy)Cl]$^+$, eq 7. The challenge in applying this method to· ultrasmall electrodes is that values of Q associated with the adsorbed redox molecules are exceedingly small.

$$A = Q/(nF\Gamma) \tag{7}$$

Figure 5 shows an example of the measurement. From the steady-state limiting current measured in the 2 mM FcTMA$^+$/0.2 M KCl solution, r_{app} for this electrode was determined to be 66 nm. The voltammetric response of the same electrode for the oxidation of adsorbed [Os(bpy)$_2$(dipy)Cl]$^+$ at a scan rate of 1000 V s^{-1} is also shown. The faradaic current associated with the adsorbed complex is barely visible on the large capacitive current. The position of the voltammetric peaks, however, is consistent with oxidation of the [Os(bpy)$_2$(dipy)Cl]$^+$ complex. Integration of the voltammetric waves yields an electrical charge $(1.1 \times 10^{-15}$ C) roughly equivalent to 7,000 molecules or ~11 zeptomoles.

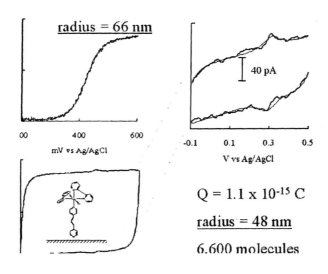

Figure 5. *(Top left) Voltammetric response (10 mVs^{-1}) of a Pt electrode (r_{app} = 66 nm) in an aqueous 2 mM FcTMA$^+$/0.2 M KCl solution. (Bottom left) Cyclic voltammogram (1000 Vs^{-1}, 200 waveforms averaged) for the same electrode corresponding to the oxidation and re-reduction of adsorbed [Os(bpy)$_2$(dipy)Cl]$^+$ in 0.2 M NaClO$_4$. (Top right) Voltammetric response plotted at higher current sensitivity. The smoothed line is the theoretical prediction based on r = 48 nm and a closed-packed molecular film of [Os(bpy)$_2$(dipy)Cl]$^+$. (Reproduced from reference 4. Copyright 2003 American Chemical Society)*

On the basis of the saturation coverage reported by Abruña and coworkers *(12)*, this charge corresponds to an electrode radius of 48 nm, roughly 70% of the value obtained from the steady-state limiting current measurement. We anticipate that it will be possible to measure as few as 500 molecules by reduction of the instrument noise.

Acknowledgements

The authors acknowledge contributions to this work from E. Maisonhaute, C. Amatore (École Normale Supérieure), H. D. Abruña (Cornell University), J. V. Macpherson and C. E. Gardner (University of Warwick) and J. Chen (Fukui University)—detailed articles with these coauthors are cited below. Financial support was provided by the Office of Naval Research, and the DOE Center for Excellence for the Synthesis and Processing of Advanced Materials, the *Science of Localized Corrosion*, sponsored by the Division of Materials Sciences.

References

1. Boxley, C. J.; Watkins, J. J.; White, H. S. Al_2O_3 film dissolution in aqueous chloride solutions. *Electrochem. Solid-State Lett.* **2003**, *6*, B38–B41.
2. Lee, H. C.; Xu, F.; Jeffcoate, C. S.; Isaacs, H. S. Cyclic polarization behavior of aluminum oxide films in near neutral solutions. *Electrochem. Solid-State Lett.* **2001**, *4*, B31–B34.
3. Boxley, C. J.; White, H. S.; Gardner, C. E.; Macpherson, J. V. Nanoscale imaging of the electronic conductivity of the native oxide film on titanium using conducting atomic force microscopy. *J. Phys. Chem. B.* **2003**, *107*, 9677–9680.
4. Watkins, J. J.; Chen, J.; White, H. S.; Abruña, H. D.; Maisonhaute, E.; Amatore, C. Zeptomole voltammetric detection and electron-transfer rate measurements using platinum electrodes of nanometer dimensions. *Anal. Chem.* **2003**, *75*, 3962–3971.
5. Conyers, J. L.; White, H. S. Electrochemical characterization of electrodes with submicrometer dimensions. *Anal. Chem.* **2000**, *72*, 4441–4446.
6. Katemann, B. B.; Schuhmann, W. Fabrication and characterization of needle-type Pt-disk nanoelectrodes. *Electroanal.* **2002**, *14*, 22–28.
7. *(a)* Morris, R. B.; Franta, D. J.; White, H. S. Electrochemistry at Pt band electrodes of width approaching molecular dimensions–breakdown of transport equations at very small electrodes. *J. Phys. Chem.* **1987**, *91*, 3559–3564; *(b)* Smith, C. P.; White, H. S. Theory of the voltammetric response of electrodes of submicron dimensions—violation of electroneutrality in the

presence of excess supporting electrolyte. *Anal. Chem.* **1993**, *65*, 3343–3353.

8. Fan, F. R. F.; Bard, A. J. Electrochemical detection of single molecules. *Science* **1995**, *267*, 871–874.

9. Bach, C. E.; Nichols, R. J.; Beckmann, W.; Meyer, H.; Schulte, A.; Besenhard, J. O.; Jannakoudakis, P. D. Effective insulation of scanning-tunneling-microscopy tips for electrochemical studies using an electropainting method. *J. Electrochem. Soc.* **1993**, *140*, 1281–1284.

10. Schulte, A.; Chow, R. H. A simple method for insulating carbon fiber microelectrodes using anodic electrophoretic deposition of paint. *Anal. Chem.* **1996**, *68*, 3054–3058.

11. Slevin, C. J.; Gray, N. J.; Macpherson, J. V.; Webb, M. A.; Unwin, P. R. Fabrication and characterisation of nanometre-sized platinum electrodes for voltammetric analysis and imaging. *Electrochem. Commun.* **1999**, *1*, 282–288.

12. Acevedo, D.; Abruña, H. D. Electron-transfer study and solvent effects on the formal potential of a redox-active self-assembling monolayer. *J. Phys. Chem.* **1991**, *95*, 9590–9594.

Chapter 46

Using Mesoporous Nanoarchitectures to Improve Battery Performance

Bruce Dunn*, François Bonet, Liam Noailles, and Paul Tang

Department of Materials Science and Engineering, University of California
at Los Angeles, Los Angeles, CA 90095–1595
*Corresponding author: email: bdunn@ucla.edu; fax: 1–310–206–7353

Aerogels are composed of a three-dimensional network of nanometer-sized solid particles surrounded by a continuous macroporous and mesoporous volume. The porosity provides both molecular accessibility and rapid mass transport and for these reasons, aerogels have been widely used in the heterogeneous catalytic materials field *(1)*. However, electrochemical materials in general, and battery materials in particular, have yet to exploit this nanoarchitecture, despite the fact that one would expect such physical features to be desirable for electrochemical reactions *(2)*. This paper reviews some of the interesting, and unexpected, electrochemical results obtained with vanadium oxide aerogels. These results underscore the benefits to be gained by creating battery materials with mesoporous architectures.

Aerogels differ substantially from traditional lithium intercalation electrode materials. First, aerogels are nanocrystalline, if not totally amorphous materials. The solid phase in an aerogel is composed of a colloidal oxide network that is characteristically quite thin, usually on the order of 10 to 50 nm. One can expect that diffusion distances will be rather short compared to traditional electrode materials. The high surface area of aerogels is significant because it means that surface effects can be amplified. Thus, surface defects that may not be evident in bulk materials, can now become prominent in aerogels because of the drastic increase in surface area. Moreover, since oxides have inherently defective surfaces, the defect chemistry associated with high vacancy concentrations is likely to influence aerogel properties. Finally, the high porosity and mesoscopic pore diameters of aerogel structures enable the electrolyte to penetrate the entire aerogel particle.

Electrochemical Properties of Vanadium Oxide Aerogels

In this paper, we briefly describe two studies that show how the mesoporous architecture of vanadium oxide aerogels leads to unique electrochemical properties. Early investigations of the electrochemical properties of aerogels did not fully appreciate the importance of preserving aerogel morphology in the electrode structures required for determining electrochemical properties. That is, in traditional electrode structures, where one combines the aerogel with other components: carbon black conductor, polymer binder and solvent, the aerogel morphology could be compromised. If the wrong solvent is used, the aerogel particles tend to agglomerate during the processing of the traditional electrodes, which reduces their surface area and collapses the interconnected mesoporous network. Moreover the carbon black particles aggregate and may occlude the aerogel surface. For these reasons, the electrochemical responses reported in many earlier studies did not represent the electrochemical properties of a high-surface-area aerogel, but rather the behavior of an agglomerated aerogel system, which did not necessarily possess an interconnected mesoporous architecture.

In our recent research, we have deliberately created electrode structures that preserve aerogel morphology. One approach has been to examine the fundamental electrochemical properties of the V_2O_5 aerogel by using "sticky-carbon" electrodes *(3)*. The aerogel particles are not aggregated on the electrode and there is ample electron, ion and solvent transport to the solid phase of the V_2O_5. The voltammetric response for the aerogel immobilized on the sticky-carbon electrode is shown in Figure 1 *(4)*. In the $LiClO_4$/propylene carbonate (PC) electrolyte, the features are broad and capacitive and the intercalation peaks appear superimposed upon the capacitive response. This capacitive response is substantially different from that which is obtained for the same V_2O_5 aerogel when it is prepared in a conventional composite electrode structure (i.e., with carbon black as a conductive additive). In that case, the voltammetric response consists of faradaic peaks associated with the characteristic intercalation behavior for sol–gel-derived V_2O_5 materials *(4)*.

The electrochemical responses for vanadium oxide aerogels can be described in terms of both a specific capacitance ($F g^{-1}$) and as lithium capacity ($mA h g^{-1}$). Depending upon the nature of the drying process, the specific capacitance values range from 960 $F g^{-1}$ to over 2000 $F g^{-1}$. The largest specific capacitance was obtained using cyclohexane as the drying solvent (2150 $F g^{-1}$), which corresponds to nearly 1500 $\mu F cm^{-2}$. This magnitude of specific capacitance represents pseudocapacitive behavior *(5)*, which is a very different charge-storage mechanism than that which occurs with traditional intercalation materials *(2)*. At the same time, these materials also possess very high capacity for lithium. Values range from 400 mAh g^{-1} to over 600 mAh g^{-1}, which are many times larger than the 140 mAh g^{-1} range exhibited by the cathodes used in commercial lithium secondary batteries *(6)*.

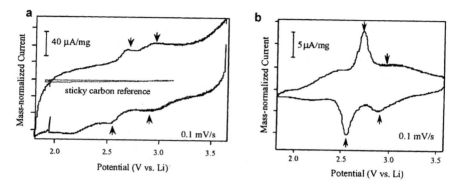

Figure 1. *Voltammograms for V$_2$O$_5$ aerogels using different electrodes: (a) Sticky carbon electrode. (b) Traditional electrode structure using carbon black as the conductive additive and PVDF as a binder. The arrows in (a) and (b) refer to the same voltages. The electrolyte is LiClO$_4$ (1 M) in propylene carbonate. (Reproduced from reference 4. Copyright 2000 Electrochemical Society.)*

The sticky-carbon electrode studies show that the inherent electrochemical properties of the aerogel are different from those obtained using a traditional composite electrode. In this electrode, the entire particle is electrochemically accessible, with much shorter diffusion distances than those that occur in traditional electrodes that involve micron-scale dimensions for electron and ion transport. Under these conditions, the V$_2$O$_5$ aerogel displays a combination of battery-like ion incorporation and capacitor-like response. That is, the V$_2$O$_5$ aerogels possess a high specific capacitance, with values comparable to those of supercapacitors, as well as a reversible capacity for lithium that is substantially greater than that of commercial intercalation materials. We have proposed that the origin of this behavior is associated with the high surface area of the aerogel, meaning that the aerogel morphology amplifies surface phenomena *(2)*. Surface defects, such as cation vacancies, which are not prominent in bulk materials, are expected to be present in high concentrations because of the high surface area of the aerogel. Lithium ion access to such defect sites provides charge compensation for the vacancies and explains how higher lithium capacities are achieved in aerogels as compared to xerogels and crystalline V$_2$O$_5$ *(6)*. Recent research by Swider-Lyons et al. provides greater insight concerning the role of defect structure on lithium capacity *(7)*.

The second type of electrode structure that preserves aerogel morphology is based on using carbon nanotubes as the electronically conducting network in V_2O_5 aerogel electrodes. Battery electrodes require the addition of an electronically conducting phase to overcome the low electronic conductivity of transition metal oxides and facilitate electrochemical reactions *(6)*. Although carbon black is the typical additive, the unique morphology of vanadium oxide aerogels raises the question of whether carbon black is the best conductive phase for this system. In contrast to carbon black, which can aggregate and impede electrolyte access, carbon nanotubes possess a similar morphology and dimensional scale as the vanadium oxide ribbons that compose the aerogel. This morphological resemblance suggested the prospect of fabricating a nanocomposite that exploits the high conductivity of SWNTs and does not block the high surface area of the aerogel. In view of the capacitor-like behavior exhibited by the aerogel using sticky carbon electrodes, it was anticipated that the V_2O_5-SWNT electrodes would have good performance at high discharge rates.

We developed a synthetic approach whereby gelation of vanadium oxide occurred around the SWNTs *(8)*. The resulting nanocomposites exhibit intimate contact, at the nanodimensional level, between the nanotubes and the vanadium oxide ribbons that compose the aerogel. This morphology leads to excellent charge transfer properties between the two phases as shown by galvanostatic studies carried out for different composite electrodes. The results (Figure 2) show that the V_2O_5–SWNT electrodes exhibit excellent electrochemical properties, particularly at high discharge rates *(8)*. Above discharge rates of 2 C, it is evident that the specific capacities for the V_2O_5–SWNT electrodes are consistently higher than that of the traditional electrode even though the V_2O_5–SWNT electrodes contain about half as much carbon (9 wt.% vs. 17 wt.%). The decrease in lithium capacity at high discharge rates is much less prominent with the V_2O_5–SWNT electrodes than with standard electrode structures.

In summary, our studies show that aerogels possess a novel and very desirable microstructure for electrochemical systems. Diffusion lengths are short and the mesoporous morphology enables the electrolyte to penetrate the entire aerogel particle. Among the more significant properties exhibited by these materials are a substantial increase in lithium capacity at high discharge rates and fundamental electrochemical behavior which combines both battery and capacitor characteristics. It is believed that the unique properties of aerogels are influenced significantly by the presence of defects, which reach appreciable concentrations in high-surface-area aerogels and now dominate the electrochemical behavior.

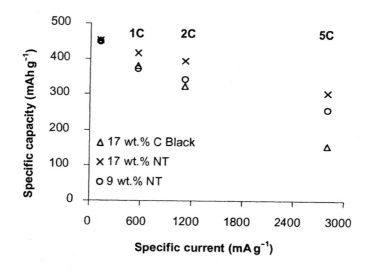

Figure 2. *Specific capacity as a function of discharge rate (to 5 C) for V_2O_5–SWNT nanocomposite electrodes with 9 and 17 wt. % SWNT compared to a traditional electrode (V_2O_5–Carbon Black), which had an equivalent amount of carbon black additive (17 wt. %). (Reproduced from reference 8. Copyright 2002 Electrochemical Society.)*

Acknowledgements

The authors are grateful for the support of their research by the Office of Naval Research.

References

1. Pajonk, G. M. Catalytic aerogels. *Catal. Today* **1997**, *35*, 319–337.
2. Rolison, D. R.; Dunn, B. Electrically conductive oxide aerogels: new materials in electrochemistry. *J. Mater. Chem.* **2001**, *11*, 963–980.
3. Long, J. W.; Swider, K. E.; Merzbacher, C. I.; Rolison, D. R. Voltammetric characterization of ruthenium oxide-based aerogels and other RuO_2 solids: the nature of capacitance in nanostructured materials. *Langmuir* **1999**, *15*, 780–785 [correction: *Langmuir* **2003**, *19*, 2532].

4. Dong, W.; Rolison, D. R.; Dunn, B. Electrochemical properties of high surface area vanadium oxide aerogels. *Electrochem. Solid-State Lett.* **2000,** *3,* 457–459.

5. Conway, B. E. Transition from supercapacitor to battery behavior in electrochemical energy-storage. *J. Electrochem. Soc.* **1991,** *138,* 1539–1548.

6. Winter, M.; Besenhard, J. O.; Spahr, M. E.; Novák, P. Insertion electrode materials for rechargeable lithium batteries. *Adv. Mater.* **1998,** *10,* 725–763.

7. Swider-Lyons, K. E.; Love, C. T.; Rolison, D. R. Improved lithium capacity of defective V_2O_5 materials. *Solid State Ionics* **2002,** *152-153,* 99–104.

8. Sakamoto, J. S.; Dunn, B. Vanadium oxide–carbon nanotube composite electrodes for use in secondary lithium batteries. *J. Electrochem. Soc.* **2002** *149,* A26–A30.

Chapter 47

Nanostructured Superlattice Thin-Film Thermoelectric Devices

Rama Venkatasubramanian*, Edward Siivola, Brooks O'Quinn, Kip Coonley, Thomas Colpitts, Pratima Addepalli, Mary Napier, and Michael Mantini

Research Triangle Institute, Research Triangle Park, NC 27709
*Corresponding author: email: rama@rti.org; fax: 1–919–541–6515

Introduction

Thin-film nanostructured materials offer the potential to dramatically enhance the performance of thermoelectrics, thereby offering new capabilities ranging from efficient cooling of small footprint communication lasers to eliminating hot spots in microprocessor chips in the near-term, to CFC-free refrigeration, portable electric power sources for replacing batteries, thermochemistry-on-a-chip, etc. in the long-term. We demonstrated *(1)* a significant enhancement in the thermoelectric figure-of-merit (ZT) at 300K. We have shown a ZT of about 2.4 in 1-nm/5-nm p-type Bi_2Te_3/Sb_2Te_3 superlattice structures and recently, of about 1.7 to 1.9 in 1-nm/4-nm n-type $Bi_2Te_3/Bi_2Te_{3-x}Se_x$ superlattices. These improvements have been realized using the concept of phonon blocking, electron-transmitting superlattice structures. The phonon blocking arises from a complex localization-like behavior for phonons in nanostructured superlattices and the electron transmission is facilitated by optimal choice of band-offsets in these semiconductor heterostructures. This represents a significant improvement over state-of-the-art ZT of ~0.9 to 1 in bulk Bi_2Te_3-based thermoelectric materials. The concept of using superlattices to obtain enhanced ZT over alloys has been demonstrated in other material systems as well, including achieving a ZT of ~0.8 in Si/Ge superlattices at 300 K, as compared to ~0.1 in SiGe alloys *(2)*. Note that although this represents an eightfold improvement and utilizes Si-based materials, even this magnitude of ZT is insufficiently attractive for widespread applications. More recently Harman et al. *(3)* have demonstrated a ZT of about 1.6 in PbTe/PbTeSe quantum-dot superlattices and Beyer et al. *(4)* have indicated a ZT of ~0.8 at 300 K in PbTe/Bi$_2$Te$_3$ superlattices, compared to a ZT of ~0.5 in PbTe-based alloys. Most of these successful demonstrations have been based on the concept of reduction in thermal conductivity with nanostructures *(5–8)*.

The thin-film devices, resulting from microelectronic processing, allow high cooling power densities to be achieved for a variety of applications, with potential localized active-cooling power densities approaching 700 Wcm^{-2}. In addition to high performance and power densities, these thin-film microdevices are also extremely fast acting, with time constants of ~10 to 20 μsec, which is about a factor of 23,000 faster than bulk thermoelectric devices. These early results *(1)* have set the stage for a wide range of applications for the superlattice thin-film thermoelectric technology.

This paper focuses on our transitioning the enhanced figure-of-merit (ZT) in p-type Bi$_2$Te$_3$/Sb$_2$Te$_3$ and n-type Bi$_2$Te$_3$/Bi$_2$Te$_{3-x}$Se$_x$ superlattices *(1)* to demonstrable functionality and performance at the module level with several initial device demonstrations. Although we have made significant headway, there remains much to be done to realize the full potential impact of the intrinsic ZT of superlattice materials.

Experimental Methods and Results

The p-type Bi$_2$Te$_3$/Sb$_2$Te$_3$ and n-type Bi$_2$Te$_3$/Bi$_2$Te$_{3-x}$Se$_x$ superlattices were grown by a low-temperature metallorganic chemical vapor deposition technique *(9,10)* that allows the deposition of high-quality superlattice interfaces in "low-temperature materials," with periodic van der Waals bonding along the growth direction. This low-temperature growth process has been indicated *(11)* to be essential in obtaining the enhanced ZT in the Bi$_2$Te$_3$/Sb$_2$Te$_3$ and n-type Bi$_2$Te$_3$/Bi$_2$Te$_{3-x}$Se$_x$ superlattice materials.

The fundamental cooling or power conversion unit in an operational thermoelectric module is the p–n couple. An important next step towards device technology has been to demonstrate working p–n couples using these thin-film superlattice materials. We note that a full device implementation with thin-films in general and using these high-performance materials, in particular, offers many significant challenges. One involves the minimization of ohmic resistances to the p- and n-thermoelectric elements, which was already discussed in Ref. 1, as part of the extrinsic ZT demonstration. The next step is to minimize the interconnect resistances between the p- and n-type elements. Towards this end, we have been able to obtain a best ZT of ~2 in a superlattice p–n couple *(12)*.

With these early p–n couples, we have looked at both cooling and power conversion device demonstrations. Shown in Figure 1 is the schematic of a p–n couple in a power-conversion test mode. The p–n couple exhibited a ZT of as much as 1.6, as obtained from a heat-to-power efficiency method *(13)*. Single p–n couples have produced a power of ~8.5 mW for a ΔT across the couple of 29 K. These devices indicate an active (as defined by the power produced by actual area of the p–n couple) power density of ~24 W cm^{-2}. When the thermal interface resistances between the active device and the external heat source were optimized, as much as 85 % of the external ΔT was transferred to the device. A power level of 38 mW per couple has been obtained with a 4-μm-thick element at a ΔT of about 107 K. This performance translates to an active power density of ~54 W cm^{-2} and a mini-module power density of ~10.5 W cm^{-2} *(12)*.

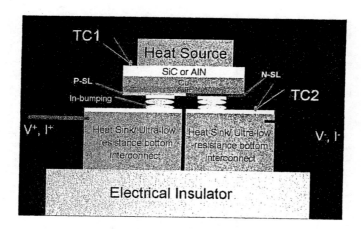

Figure 1. Schematic of a thin-film superlattice p–n couple in a power conversion test mode showing the p-type superlattice (P-SL) and n-type superlattice (N-SL), where TC1 and TC2 are thermocouples and V, I represent the voltage and currents at the positive and negative termini.

In the cooling demonstration effort, we have been able to obtain over 40 K active cooling with thin-film superlattice cooling devices, useable in several laser and microprocessor cooling needs where the hot-sides run at ~80 °C. This level of performance is demonstrated in the cooling data shown in Figure 2. We note that this cooling has been achieved in spite of severe thermal management issues that had to be overcome, and are not yet completely resolved. Because of to insufficient thermal management at high heat-flux levels, the "true" hot-side temperature and hence the "true" ΔT across the device is much higher than observed in Figure 2. Even so, these small-scale, 600 μm × 600 μm superlattice thermoelectric cooling modules can address hot-spot cooling needs of current and future generations of microprocessors *(14)*.

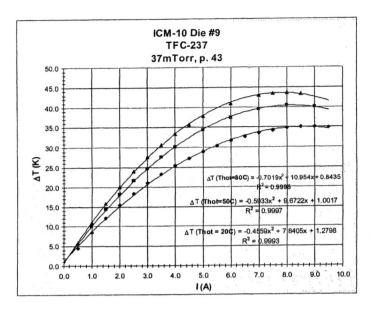

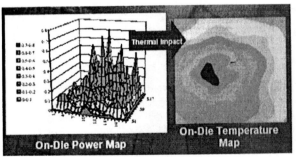

Figure 2. The preliminary cooling performance in a 600 μm × 600 μm superlattice thermoelectric cooling module that can address the emerging spot-cooling needs of microprocessors (reproduced from reference 14) and other communication laser devices.

Conclusion

The transition from the high "intrinsic" ZT attainable in thin-film p-type Bi_2Te_3/Sb_2Te_3 and n-type $Bi_2Te_3/Bi_2Te_{3-x}Se_x$ superlattice materials and the high "extrinsic" ZT attainable in their respective thermoelements (1) to module-level performance is a huge engineering challenge. However, such a transition would enable a plethora of possibilities ranging from thermal management of hot spots in microprocessors to efficient thermal-to-electric power conversion devices. With the results presented here, both power and cooling thermoelectric devices with thin-film superlattice materials appear to be near-term potential realities. Some of the challenges that remain to be addressed in the full development of this nanoscale thermoelectric-materials technology include optimization of the various electrical and thermal interfaces both within the p–n couple and between the couple and the external thermal management components.

Acknowledgements

The thermoelectric cooling results presented here were made possible through support by the Defense Advanced Research Projects Agency (DARPA) and the Office of Naval Research (ONR) under U. S. Navy Contract No. N00014-97-C-0211. The thermoelectric power conversion results presented here were supported by DARPA through an Army Research Office (ARO) Contract No. DAAD19-01-C-0070. These supports are gratefully acknowledged.

References

1. Venkatasubramanian, R.; Siivola, E.; Colpitts, T.; O'Quinn, B. Thin-film thermoelectric devices with high room-temperature figures of merit. *Nature* **2001**, *413*, 597–602.
2. Venkatasubramanian, R.; Siivola, E.; Colpitts, T. S. *Proc. of 17th International Conference on Thermoelectrics*, IEEE Catalog No. 98TH8365, 1998, p 191.
3. Harman, T.; Taylor, P. J.; Walsh, M. P.; LaForge, B. E. Quantum dot superlattice thermoelectric materials and devices. *Science* **2002**, *297*, 2229–2232.
4. Beyer, H.; Nurnus, J.; Bottner, H.; Lambrecht, A.; Wagner, E.; Bauer, G. High thermoelectric figure of merit ZT in PbTe and Bi_2Te_3-based superlattices by a reduction of the thermal conductivity. *Physica E*, **2002**, *13*, 965–968.
5. Venkatasubramanian, R.; Timmons, M. L.; Hutchby, J. A.; Borrego, J. *Proc. of 1st National Thermogenic Cooler Workshop*; Horn, S., Ed.; Fort Belvoir, VA, 1992, pp 196–231.
6. Venkatasubramanian, R.; Timmons, M. L.; Hutchby, J. A. *Proc. of 12th International Conference on Thermoelectrics*, Yokohama, Matsuura, K., Ed.; 1993, p 322.

7. Lee, S. M.; Cahill, D. G.; Venkatasubramanian, R. Thermal conductivity of Si–Ge superlattices. *Appl. Phys. Lett.* **1997**, *70*, 2957–2959.
8. Venkatasubramanian, R. Lattice thermal conductivity reduction and phonon localization-like behavior in superlattice structures. *Phys. Rev. B* **2000**, *61*, 3091–3097.
9. Venkatasubramanian, R.; Colpitts, T.; O'Quinn, B.; Liu, S.; El-Masry, N.; Lamvik, M. Low-temperature organometallic epitaxy and its application to superlattice structures in thermoelectrics. *Appl. Phys. Lett.* **1999**, *75*, 1104–1106.
10. Venkatasubramanian, R. Low temperature chemical vapor deposition and etching apparatus and method. U.S. Patent no. 6,071,351 (6 June 2000).
11. Venkatasubramanian, R.; Siivola, E.; Colpitts, T.; O'Quinn, B. Phonon-blocking, electron-transmitting low-dimensional structures. U.S. Patent Application No. 20,030,099,279, www.uspto.gov.
12. Venkatasubramanian, R.; Siivola, E.; O'Quinn, B. C.; Coonley, K.; Addepalli, P.; Napier, M.; Colpitts, T.; Mantini, M., to be published.
13. Goldsmid, H. J. *Electronic Refrigeration*, Pion Ltd., 1983.
14. Bannerjee, K.; Mahajan, R. www.intel.com/showcase/silicon.

Chapter 48

Membraneless Fuel Cells: An Application of Microfluidics

Tzy-Jiun M. Luo, Jiangfeng Fei, Keng G. Lim,
and G. Tayhas R. Palmore*

Division of Engineering, Brown University, Providence, RI 02912
*Corresponding author: Tayhas_Palmore@brown.edu

Introduction

Research activity in fuel-cell technology has increased remarkably in the past few years primarily because of environmental concerns. For example in the area of transportation, fuel cells promise to increase the efficiency of converting fossil fuels into useful work (i.e., transportation) while reducing the amount of toxins released into the environment. In the area of portable electronics, fuel cells offer an attractive alternative to batteries and the associated environmental impact of their disposal (1).

Secondary to reducing the environmental impact of current energy conversion technology, is the growing need to power miniaturized microelectronic, micromechanical and microfluidic devices, such as those used in environmental sensors and medical implants and their transmitters and receivers. Currently, the lower limit in terms of size and weight of these types of miniaturized devices is determined by the volume and weight of its battery. For example, lithium-ion batteries power many medical implants, including pacemakers, atrial defibrillators, neurostimulators, and drug infusion pumps. As the size of a lithium-ion battery is reduced, its energy capacity does not scale with its volume, because the volume fraction of the casing, seals and current collectors increase. Thus, further reduction in the size of medical implants and the emerging "lab-on-a-chip" technology will require the development of alternatives to battery technology.

Background

A fuel cell is an electrochemical device that manages the flow of electrons and charge-compensating positive ions in a redox reaction in a manner that moves the electrons through an external circuit where they can do electrochemical work (Figure 1). The fuels used in a typical low-temperature fuel cell are H_2 and O_2; other reducing fuels such as methanol often are converted to H_2 before entering the fuel cell. Generation of power occurs upon addition of fuel to the fuel cell.

Figure 1. *A dihydrogen/dioxygen fuel cell illustrating the direction of electron and ion flow. When the circuit is closed, electrons flow from the anode to the cathode through the external circuit and the positively charged ions (i.e., protons) flow through the membrane from the anode compartment to the cathode compartment.*

A biofuel cell is a combination of two technologies: fuel cells and biotechnology. Similar to conventional fuel cells, biofuel cells consist of an anode and cathode separated by a barrier that is selective for the passage of positively charged ions *(2,3)*. Unlike conventional fuel cells, which use precious metals as catalysts, biofuel cells utilize enzymatic catalysts, either as they occur in microorganisms, or more commonly, as isolated proteins. The use of microorganisms or enzymatic catalysts instead of noble metal catalysts results in a fuel cell that functions under mild working conditions: ambient temperature and pressure, and moderate pH. Consequently, biofuel cells are attractive devices for powering miniaturized microelectronic, micromechanical and microfluidic components in medical implants or environmental sensors.

In the case of medical implants, the accompanying miniature fuel cell would be fueled by glucose and oxygen delivered via the circulatory system. Several prototype fuel cells recently have been reported that use glucose oxidase as the anodic catalyst and laccase or bilirubin oxidase as the cathodic catalyst *(4-9)*. The generation of electrical power by these prototypes requires either a membrane to separate the fuels or, if absent, selective electrocatalysis. Difficulties in manufacturing a miniature fuel cell based on either of these configurations, however, make alternative designs desirable.

One possible design exploits laminar flow of fluids at low Reynolds number. The Reynolds number is a dimensionless parameter that relates the ratio of inertial forces to viscous forces of a specific configuration of fluid flow. Specifically, the Reynolds number (Re) is given by eq 1:

$$Re = vl\rho\eta^{-1} \tag{1}$$

where v is velocity in $cm\ s^{-1}$, l is the diameter of the channel, ρ is the density of the liquid in $g\ cm^{-3}$, and η is the dynamic viscosity in $g\ cm^{-1}s^{-1}$. In our case, the value for each parameter in eq 1 is $v = 1\ cm\ s^{-1}$, $l = 50\ \mu m$ and is the height of the channel, and approximating the fluid to be pure water, $\rho = 1\ g\ cm^{-3}$ and $\eta = 8.94 \times 10^{-3}\ g\ cm^{-1}s^{-1}$. Under these conditions, $Re = 0.56$ and fluid flow is predicted to be laminar. It is this flow behavior that a fuel cell designed around a microfluidic channel exploits.

Two fluids such as the anolyte and catholyte in a microfluidic fuel cell can flow next to each other in direct contact (i.e., liquid–liquid interface) and not mix by turbulence. Instead, mixing occurs only by diffusion of molecules across the interface. This type of design removes the need for either a membrane or selective electrocatalysts. Depending on the location of the electrodes relative to the fluid–fluid interface, diffusion of molecules across the interface may affect the cell potential and thus, power output of the fuel cell.

Materials and Methods

In our design, the fuel cell consists of a central channel 300- to 800-μm wide and 50 μm in height. The channels are lined with electrode pairs measuring 300 μm by 4 mm, with the electrodes in each pair separated by 200 μm. Fuel can be delivered to the electrode array within the channel via the inlet ports and discharged (or recycled) via the outlet ports. Diffusive transport of molecules from one electrode to the other in each electrode pair is approximated to be in the range of milliseconds to seconds depending on molecular size.

Microfluidic channels were fabricated using replica-molding techniques *(10,11)*. Briefly, a master channel is made in photoresist (Micochem SU-850, 50-μm high, 800-μm wide) by photolithography using a mask fabricated from a high-resolution transparency. This negative-relief master is replicated by molding in PDMS, which is peeled from the substrate to produce a slab of PDMS containing channels ~50 μm in depth and of varying width depending on the configuration of the electrode array and the desired pattern of flow.

A metal-lift off procedure was used to fabricate microelectrode arrays with different configurations *(12)*. This procedure is inexpensive and facile, thus enabling the rapid-prototyping of new electrode configurations on the basis of new results. Briefly, the patterns for the electrodes and electrical leads are generated initially with CAD software and printed at high resolution onto an overhead transparency to generate an inexpensive mask for photolithography. Glass slides (or other substrates) are spin-coated with Shipley 1813 photoresist to a desired thickness (i.e., typically 1–2 μm), exposed to ultraviolet light through the overlaying mask, and developed in tetramethylammonium hydroxide. Subsequent to pattern formation on a substrate, a conductor is evaporated to a thickness of ~1000 Å on top of the patterned photoresist. Dissolving the underlying photoresist with acetone facilitates the lift-off of the unwanted segments of conductor to obtain the desired microelectrode array. Covering the microelectrode array with the channel-containing PDMS slab completes the fabrication of the microfluidic fuel cell.

Results

Shown in Figure 2a is a photograph of a prototype microfluidic biofuel cell displaying the microelectrode array enclosed by the central channel of the PDMS slab. Electrical contact to the enclosed electrodes is made via the gold lines that lead to large gold pads external to the PDMS slab. The dashed box indicates the section of the device magnified and shown in Figure 2b, which reveals the laminar flow of fuel and oxidant.

Shown in Figure 3a are the molecular structures of the anodic fuel (1,4-dihydrobenzoquinone, BQH_2) and cathodic mediator {2,2'-azinobis (3-ethyl-benzothiazoline-6-sulfonate), ABTS} used in the microfluidic fuel cell shown in Figure 2. Two electrons and two protons are liberated at the anode upon oxidation of each molecule of BQH2 to BQ (Figure 3b). Electrons are transported via an external circuit to become available at the cathode while protons diffuse across the fluid–fluid interface for charge balance. Bioelectrocatalytic reduction of dioxygen in the catholyte was achieved with the copper oxidase, laccase, and the cathodic mediator, ABTS *(13)*.

Shown in Figure 3c is a plot of voltage as a function of current density generated by this device as fueled with 1,4-dihydrobenzoquinone (anodic fuel) and dioxygen as the terminal oxidant. The performance of the fuel cell, reflected by the shape of its *current density–voltage* curve, is similar to that

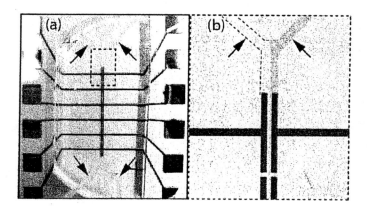

Figure 2. *(a) A microfluidic biofuel cell consisting of a linear channel that encloses six pairs of electrodes. The location of ports for entry and exit of fuel and oxidant are indicated by the arrows. (b) Magnified view of area enclosed in the dashed box in (a).—Arrows point to fuel (dark gray on right inlet) and oxidant (outlined on left inlet as color is indistinguishable from background color). Only the first pair and part of the second pair of electrodes are seen in this photo (black lines).*

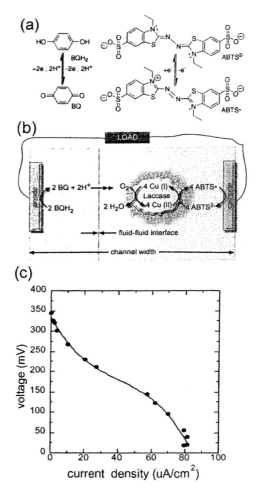

Figure 3. (a) Molecular structure of anodic fuel and cathodic mediator used in the microfluidic fuel cell. (b) Electron and proton flow during bioelectrocatalytic reduction of dioxygen. (c) Current density vs. voltage plot of the microfluidic biofuel cell operated at different loads.

observed with fuel cells that contain a membrane separating catholyte and anolyte. The open-circuit voltage of the device was 350 mV, reflecting the potential difference between 1,2-dihydrobenzoquinone and ABTS. Maximum current density for this prototype was 80 μA cm^{-2} and the maximum power density was 8.3 μW cm^{-2} at 150 mV.

Two other microfluidic fuel cells recently were reported in the literature. In the first example, few specifics were given as to the electrocatalyst or anodic fuel used (14). Dioxygen was the terminal oxidant and both anodic fuel and dioxygen were delivered to the anode and cathode, respectively, as aqueous solutions. The maximum power density was $0.22 \, mW \, cm^{-2}$ at $0.45 \, V$, corresponding to a current density of $\sim 0.5 \, mA \, cm^{-2}$. In the second example, 1 M solutions of vanadium ions dissolved in 25 % H_2SO_4 were used as anodic [V(III)/V(II)] and cathodic fuels [V(V)/V(IV)] (*15*). The electrodes in this device were carbon-coated gold and the open-circuit voltage was $1.59 \, V$. The maximum current density was $35 \, mA \, cm^{-2}$ at $1.1 \, V$, corresponding to $38 \, mW \, cm^{-2}$. It is difficult to compare the performance of these two reported devices to our device for several reasons. First, different fuels and oxidants were used, which impacts the open-circuit potential of the fuel cell (e.g., $1.1 \, V$ vs. $1.59 \, V$ vs. $0.35 \, V$), and ultimately the maximum current density and power density. Second, although all examples use aqueous solutions, the pH is different in each case (e.g., unspecified vs. pH 0 vs. pH 5), which impacts the impedance of the device. Third, the height and width of the channels, respectively, are different (e.g., 1 mm × 1 mm vs. 0.2 mm × 2 mm vs. 0.050 mm × 0.8 mm). Fourth, the rate at which fuel flowed through the microfluidic fuel cell is different in each case (e.g., unspecified vs. $12.5 \, cm \, s^{-1}$ vs. $1 \, cm \, s^{-1}$). Finally, the distance between the anode and cathode is different (e.g., unspecified vs. 1 mm vs. 0.20 mm). Although the performance of these devices cannot be compared directly, power is generated by all three examples and thus provides a basis for a new type of fuel cell configuration that does not require a membrane or selective catalysts.

Conclusions

Our results demonstrate that the behavior of fluid flow at low Reynolds number can be exploited to generate small amounts of electrical power when used in combination with an electrode array and bioelectrocatalysts. Further efforts will focus on improving the performance of this type of fuel cell configuration (*i.e.*, membraneless) with other channel/electrode configurations and biocatalysts.

References

1. Dyer, C. K. Fuel cells for portable applications. *J. Power Sources* **2002**, *106*, 31–34.
2. Palmore, G. T. R.; Whitesides, G. M. Microbial and enzymatic biofuel cells. In *Enzymatic Conversion of Biomass for Fuels Production;* Himmel, M. E., Baker, J. O., Overend, P., Eds.; Am. Chem. Soc. Symp. Ser. 566: Washington, DC, 1994; pp 271–290.
3. Kreysa, G.; Sell, D.; Krämer, P. Bioelectrochemical fuel cells. *Ber. Bunsenges. Phys. Chem.* **1990**, *94*, 1042–1045.
4. Katz, E.; Bückmann, A. F.; Willner, I. Self-powered enzyme-based biosensors. *J. Am. Chem. Soc.* **2001**, *123*, 10752–10753.

5. Palmore, G. T. R.; Kim, H.-H. Electro-enzymatic reduction of dioxygen to water in the cathode compartment of a biofuel cell. *J. Electroanal. Chem.* **1999**, *464*, 110–117.

6. Katz, E.; Willner, I.; Kotlyar, A. B. A non-compartmentalized glucose|O_2 biofuel cell by bioengineered electrode surfaces. *J. Electroanal. Chem.* **1999**, *479*, 64–68.

7. Tsujimura, S.; Fujita, M; Tatsumi, H.; Kano, K.; Ikeda, T. Bioelectrocatalysis-based dihydrogen/dioxygen fuel cell operating at physiological pH. *Phys. Chem. Chem. Phys.* **2001**, *3*, 1331–1335.

8. Chen, T.; Barton, S. C.; Binyamin, G.; Gao, A.; Zhang, Y. Kim, H.-H.; Heller, A. A miniature biofuel cell. *J. Am. Chem. Soc.* **2001**, *123*, 8630–8631.

9. Mano, N.; Kim, H.-H.; Zhang, Y.; Heller, A. An oxygen cathode operating in a physiological solution. *J. Am. Chem. Soc.* **2002**, *124*, 6480–6486.

10. Xia Y. N.; Whitesides, G. M. Soft lithography. *Angew. Chem. Int. Ed. Engl.* **1998**, *37*, 551–575.

11. Duffy, D. C.; McDonald, J. C.; Schueller, O. J. A.; Whitesides, G. M. Rapid prototyping of microfluidic systems in poly(dimethylsiloxane). *Anal. Chem.* **1998**, *70*, 4974–4984.

12. Moreau, W. M. Semiconductor lithography principles, practices, and materials. In *MicroDevices Physics and Fabrication Technologies—The Physics of Micro/Nano-Fabrication*; Muray, J. J.; Brodie, I., Eds.; Plenum: New York, 1988, p 607.

13. Gelo-Pujic, M.; Kim, H.-H.; Butlin, N. G.; Palmore, G. T. R. Electrochemical studies of a truncated laccase produced in *Pichia pastoris*. *Appl. Environ. Microbiol.* **1999**, *65*, 5515–5521.

14. Choban, E. R.; Markoski, L. J.; Stoltzfus, J.; Moore, J. S.; Kenis, P. A.; Microfluidic fuel cells that lack a PEM. *Power Sources Proc.* **2002**, *40*, 317–320.

15. Ferrigno, R.; Stroock, A. D.; Clark, T. D.; Mayer, M.; Whitesides, G. M., Membraneless vanadium redox fuel cell using laminar flow. *J. Am. Chem. Soc.* **2002**, *124*, 12930–12931 (addition: *J. Am. Chem. Soc.* **2003**, *125*, 2014).

Chapter 49

Hybrid Photoelectrochemical-Fuel Cell

Linda de la Garza[1], Goojin Jeong[1], Paul A. Liddell[1], Tadashi Sotomura[2], Thomas A. Moore[1,*], Ana L. Moore[1,*], and Devens Gust[1,*]

[1]Department of Chemistry and Biochemistry, Arizona State University, Tempe, AZ 85287–1604
[2]Advanced Technology Research Laboratories, Matsushita Electric Industrial Company Ltd., Moriguchi, Osaka 570–8501, Japan
*Corresponding authors: tmoore@asu.edu; amoore@asu.edu; and gust@asu.edu

Introduction

Plants use photosynthesis to convert light energy into electrochemical potential, and ultimately into chemical potential energy stored in reduced carbon compounds. When oxidized, these compounds provide energy to plants and heterotrophic organisms. Mimicry of these processes is being investigated as a method for electricity production *(1)*. Photovoltaic devices, such as dye-sensitized photoelectrochemical solar cells, mimic the initial steps of photosynthesis. These employ nanoparticulate wide-bandgap semiconductor electrodes coated with monolayers of dyes that absorb light in the visible range and inject electrons into the conduction band of the semiconductor *(2-4)*. A complementary aspect of biomimicry involves biofuel cells that produce electricity by oxidation of biological materials produced by photosynthesis. These cells often use enzymes to catalyze the oxidation of carbohydrates and other reduced-carbon fuel, and couple the oxidation to electrical current flow between an anode and a cathode *(5-8)*.

In a new approach to biomimicry we combined a dye-sensitized nanoparticulate semiconductor photoanode with an enzyme-catalyzed biofuel cell. The hybrid cell offers several potential advantages over either a dye-sensitized photoelectrochemical cell or a biofuel cell, including the ability to produce more power *(9)*.

Methods and Results

The photoanode consists of indium-tin oxide (ITO) coated conductive glass that supports a layer of nanoparticulate tin dioxide, designated $n\text{SnO}_2$. The $n\text{SnO}_2$ is coated with a monolayer of tetra-arylporphyrin sensitizer (S) that bears a carboxylic-acid functionality necessary for binding to the oxide. The absorption spectrum of this electrode, taken in air, features typical porphyrin absorption, with Q-bands at 650, 590, 555, and 520 nm, and a Soret band at ~420 nm.

Excitation of the dye with visible light is followed by electron injection into the conduction band of the semiconductor. This process produces an oxidized dye molecule (S^{+}). In our cell S^{+} is reduced back to its initial state by electron donation from the reduced form of nicotinamide adenine dinucleotide (NADH) that is present in the aqueous buffer solution surrounding the electrode (see Figure 1).

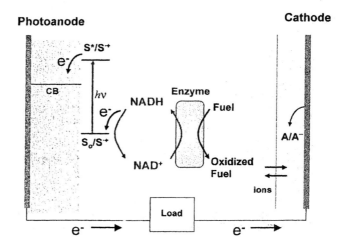

Figure 1. Schematic diagram of the hybrid cell.

NADH/NADPH are, in theory, excellent redox mediators between the photoanode and enzymes. They are redox coenzymes in many enzyme-mediated processes, have been used as electron donors for reduction of a number of oxidized organic compounds *(10-14)* and have highly reducing formal potentials that are suitable for reduction of S^{+}. However, in biofuel cells NADH/NAD^{+} has not proven to be a useful redox mediator for coupling enzymatic oxidations to the anode, due to high overpotential and electrode contamination with consequent irreproducibility. Instead, with the photoanode described above, we have observed that NADH is an excellent electron donor to $S^{\bullet+}$ yielding only NAD^+, without detectable amounts of other products. At pH = 7, the first oxidation potential of NADH is 0.73 V vs. Ag/AgCl *(15)* and it readily donates an electron to $S^{\bullet+}$ (the first oxidation potential of S is 1.03 V vs. Ag/AgCl). The second oxidation step of NADH is easier than the first (– 1.12 V), because the product of the initial one-electron oxidation, $NADH^{\bullet+}$, is highly acidic, and deprotonates to yield $NAD^{\bullet}$. This radical is short lived, and in the photoelectrochemical biofuel cell must undergo a second electron transfer to form NAD^+, as detected by NMR.

The photoelectrochemical cell was set up with a platinum gauze cathode and an electrolyte consisting of aqueous 1.0 M NaCl containing 0.25 M Tris buffer at pH 8.0. The solution contained no enzymes or biofuel. The cell was illuminated with 520-nm light at 1.0 $mW cm^{-2}$, and the photocurrent was measured. In the absence of NADH, no significant photocurrent was observed. Adding increasing amounts of NADH to the solution led to increased photocurrents, until at 3.0 mM NADH or above, a limiting photocurrent of ~35 $\mu A cm^{-2}$ was observed (see Figure 2).

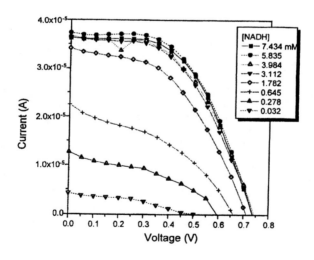

Figure 2. Current–voltage characteristic at different concentrations of NADH.

The photocurrent was measured as a function of the wavelength of the exciting light. When the current production is plotted as the incident-photon-to-current efficiency (*IPCE*), which is defined as the number of electrons in the external circuit generated by the cell divided by the number of incident photons, the *IPCE* tracks light absorption by the porphyrin in the Q-band region, indicating that light absorbed by **S** is responsible for the photocurrent (see Figure 3). These experiments clearly demonstrate that NADH can act as an electron donor to the $S^{•+}$ produced by photoinduced electron transfer to the nSnO$_2$, leading to a significant photocurrent.

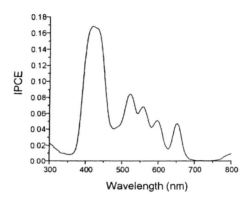

Figure 3. Photon-to-current efficiency (IPCE) as a function of wavelength.

The regeneration of NADH from NAD$^+$ produced at the photoanode is accomplished by using NAD$^+$ as coenzyme of a dehydrogenase enzyme and a reduced carbon fuel as substrate. The result is oxidation of an organic fuel and production of the NADH needed to continue electron production at the photoanode. We have focused on the use of readily available, reduced carbon fuel such as glucose, ethanol and methanol.

Methanol can be oxidized to formaldehyde by alcohol dehydrogenase (ADH), using NAD$^+$ as the electron acceptor. A second enzyme, aldehyde dehydrogenase (AlDH), uses NAD$^+$ to oxidize formaldehyde to formate, and a third, formate dehydrogenase (FDH) converts formate to carbon dioxide. Each oxidation step converts one molecule of NAD$^+$ into NADH. A cell was prepared using a two-compartment configuration with a Hg/Hg$_2$SO$_4$ cathode. The electrolyte solution contained 0.9 mM NADH. After illumination of the photoanode with 520-nm light for 20 h (1 mW cm^{-2} light intensity, 2 cm^2 active area), almost all of the NADH was consumed due to oxidation at the photoanode. Addition of 5 % by volume methanol, ADH (16.0 U mL^{-1}), AlDH (1.0 U mL^{-1}), and FDH (0.3 U mL^{-1}), with continued illumination, resulted in the return of the NADH concentration to its value prior to illumination and CO$_2$

was produced as the final product. In Figure 4 the slope of the line obtained by plotting the moles of NADH consumed vs. half of the moles of electrons going into the circuit is ∼1 (open circles). This relationship illustrates that two electrons are going into the circuit per molecule of NADH consumed. The addition of enzymes and methanol (dark circles) brings the NADH level to its initial value.

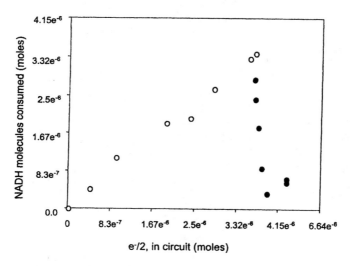

Figure 4. Moles of NADH consumed vs. half the number of moles of electrons going into the circuit (O). Level of NADH after the addition of 5 % by volume methanol, ADH, AlDH and FDH (●) (see text for details).

Discussion

These results demonstrate that it is possible to combine a dye-sensitized nanoparticulate semiconductor electrode with the components of a biofuel cell such as carbohydrates/simple alcohols and dehydrogenase enzymes. The coupling factor is the $NADH/NAD^+$ redox system which interfaces the one-electron reduction of the porphyrin radical cation produced by the photoinduced electron transfer reaction with the two-electron oxidations performed by the dehydrogenases. Although no attempts have been made to optimize the cell performance or choice of cathode, the results obtained here are encouraging, with the following parameters for the methanol cell described above: open circuit voltage of 0.75 V, short-circuit current of 80 $\mu A\ cm^{-2}$, incident-photon-to-current efficiency of 0.12 and a fill factor of 0.5 *(9).*

In principle, the hybrid cell can produce higher voltage and more power than a biofuel cell with the same fuel and cathodic half-cell due to the more negative anode potential resulting from photoinduced charge injection. Also, the hybrid cell can theoretically produce a higher voltage than a dye-sensitized photoelectrochemical cell using the same semiconductor and sensitizer, because the fuel-cell configuration allows the use of a separate cathodic half-cell (e.g., O_2 electrode). The hybrid cell voltage is determined by the potentials of the cathodic half-reaction and the conduction band/Fermi level of the semiconductor. In the dye-sensitized photoelectrochemical cell, the voltage is limited by the potentials of the redox relay (I^-/I_3^-) in the cell and the conduction band/Fermi level of the semiconductor (2,3).

The hybrid cell can produce, in principle, a higher peak current than a photoelectrochemical cell, due to rapid recycling of the NADH donor near the photoanode. After reduction of the oxidized sensitizer, in a conventional photoelectrochemical cell with I^-/I_3^- redox relay, the oxidized redox carrier must diffuse to the cathode before it can be reduced and recycled (2,3). This necessity for mass transport of the mediator is not necessary in the hybrid cell as the reduction of NAD^+ can be carried out in the immediate vicinity of the photoanode by the soluble enzyme system.

Hybrid cells are environmentally friendly and the reduced carbon fuel can come from many sources, including waste materials.

References

1. Dresselhaus, M. S.; Thomas, I. L. Alternative energy technologies. *Nature* **2001**, *414*, 332–337.
2. O'Regan, B.; Grätzel, M. A low-cost, high-efficiency solar-cell based on dye-sensitized colloidal TiO_2 films. *Nature* **1991**, *353*, 737–740.
3. Grätzel, M. Photoelectrochemical cells. *Nature* **2001**, *414*, 338–344.
4. Sauve, G.; Cass, M. E.; Coia, G.; Doig, S. J.; Lauermann, I.; Pomykal, K. E.; Lewis, N. S. Dye sensitization of nanocrystalline titanium dioxide with osmium and ruthenium polypyridyl complexes. *J. Phys. Chem. B* **2000**, *104*, 6821–6836.
5. Yahiro, A. T.; Lee, S. M.; Kimble, D. O. Bioelectrochemistry. I. Enzyme utilizing bio-fuel cell studies. *Biochim. Biophys. Acta* **1964**, *88*, 375–383.
6. Palmore, G. T. R.; Bertschy, H.; Bergens, S. H.; Whitesides, G. M. A methanol/dioxygen biofuel cell that uses NAD^+-dependent dehydrogenases as catalysts: application of an electro-enzymatic method to regenerate nicotinamide adenine dinucleotide at low overpotentials. *J. Electroanal. Chem.* **1998**, *443*, 155–161.
7. Palmore, G. T. R.; Whitesides, G. M. Microbial and enzymatic biofuel cells. In *Enzymatic Conversion of Biomass for Fuels Production*; Himmel, M. E.,

Baker, J. O., Overend, P., Eds. Am. Chem. Soc. Symp. Ser. 566: Washington, DC, 1994; pp 271–290.

8. Katz, E.; Willner, I.; Kotlyar, A. B. A non-compartmentalized glucose|O_2 biofuel cell by bioengineered electrode surfaces. *J. Electroanal. Chem.* **1999**, *479*, 64–68.

9. De la Garza, L; Jeong, G.; Liddell, P. A.; Sotomura, T.; Moore, T. A.; Moore, A. L.; Gust, D. Enzyme-based photoelectrochemical biofuel cell. *J. Phys. Chem. B* **2003**, *107*, 10252 - 10260.

10. Gorton, L. Chemically modified electrodes for the electrocatalytic oxidation of nicotinamide coenzymes. *J. Chem. Soc. Faraday Trans.* **1986**, *82*, 1245–1258.

11. Taraban, M. B.; Kruppa, A. I.; Polyakov, N. E.; Leshina, T. V.; Lusis, V.; Muceniece, D.; Duburs, G. The mechanisms of the oxidation of NADH analogs. 1. Photochemical oxidation of n-unsubstituted 1,4-dihydropyridines by various acceptors. *J. Photochem. Photobiol. A: Chem.* **1993**, *73*, 151–157.

12. Polyakov, N. E.; Taraban, M. B.; Kruppa, A. I.; Avdievich, N. I.; Mokrushin, V. V.; Schastnev, P. V.; Leshina, T. V.; Muceniece, D.; Duburs, G. The mechanisms of oxidation of NADH analogs. 3. Stimulated nuclear-polarization (SNP) and chemically induced dynamic nuclear-polarization (CIDNP) in low magnetic-fields in photooxidation reactions of 1,4-dihydropyridines with quinones. *J. Photochem. Photobiol. A: Chem.* **1993**, *74*, 75–79.

13. Martens, F. M.; Verhoeven, J. W.; Gase, R. A.; Pandit, U. K.; De Boer, T. Question of one-electron transfer in mechanism of reduction by NADH-models. *Tetrahedron* **1978**, *34*, 443–446.

14. Martens, F. M.; Verhoeven, J. W. Photoinduced electron-transfer from NADH and other 1,4-dihydronicotinamides to methyl violgen. *Rec. Trav. Chim. Pays-Bas* **1981**, *100*, 228–236.

15. Moiroux, J.; Elving, P. J. Mechanistic aspects of the electrochemical oxidation of dihydronicotinamide adenine-dinucleotide (NADH). *J. Am. Chem. Soc.* **1980**, *102*, 6533–6538.

Summary

This book brings together nanotechnology research that contributes to enhanced environmental protection directed toward human activities, in addition to helping define the problems and processes that might occur in the natural environment. With the human aspects of new technologies in mind, it covers two broad themes: the applications of nanotechnology to the environment and the implications of nanotechnology on the environment.

The applications include those aspects of nanotechnology that could be useful in dealing with present and legacy environmental problems and preventing future problems. They include treatment and remediation of existing pollutants, nanocatalysts for more selective and efficient reactions, nanotechnology-enabled green energy, metrology to measure these small materials, nanotechnology-enabled sensors for substances of environmental interest, and environmentally benign manufacturing of nanomaterials. Implications include toxicology and biointeractions of nanomaterials and nanoparticle geochemistry in water and air.

The book is a compilation of extended abstracts with introductory chapter material. It is the result of a symposium on Nanotechnology and the Environment: Applications and Implications presented from March 23–27, 2003, at the National Meeting of the American Chemical Society (ACS), sponsored by the ACS Division of Industrial and Engineering Chemistry, Inc. It is the authors' hope that this book will be useful in framing the research directions and stimulating thoughtful new questions about the relationship of this new technology and its interactions with the environment, either beneficial or potentially harmful.

Indexes

Author Index

Subject Index